# Mastering Mathematics

## HOW TO BE A GREAT MATH STUDENT

THIRD EDITION

# Mastering Mathematics

## HOW TO BE A GREAT MATH STUDENT

**THIRD EDITION**

**Richard Manning Smith, Ph.D.**
Bryant College

Australia • Brazil • Japan • Korea • Mexico • Singapore • Spain • United Kingdom • United States

BROOKS/COLE
CENGAGE Learning™

**Mastering Mathematics: How to Be a Great Math Student, Third Edition**
**Richard Manning Smith**

Editor: Robert Pirtle
Editorial Assistant: Melissa K. Duge
Manuscript Editor: Jane Townsend
Production Editor: Kirk Bomont
Production Management: Scratchgravel Publishing Services
Design Editor: E. Kelly Shoemaker
Cover Design: Roger Knox
Typesetting: Scratchgravel Publishing Services
Marketing Representative: Maureen Riopelle



For product information and technology assistance, contact us at **Cengage Learning Customer & Sales Support, 1-800-354-9706**

For permission to use material from this text or product, submit all requests online at **cengage.com/permissions**
Further permissions questions can be emailed to **permissionrequest@cengage.com**

**Library of Congress Cataloging-in-Publication Data**
Smith, Richard Manning, [date]–
Mastering mathematics : how to be a great math student / Richard Manning Smith. — 3rd ed.
p. cm.
Includes bibliographical references (p. – )
ISBN-13: 978-0-534-34947-9
ISBN-10: 0-534-34947-1
1. Mathematics—Study and teaching. I. Title.
QA11.S63 1997
510'.0711—dc21 97-21749
CIP

**Brooks/Cole**
10 Davis Drive
Belmont, CA 94002-3098
USA

Cengage Learning is a leading provider of customized learning solutions with office locations around the globe, including Singapore, the United Kingdom, Australia, Mexico, Brazil, and Japan. Locate your local office at: **international.cengage.com/region**

Cengage Learning products are represented in Canada by Nelson Education, Ltd.

For your course and learning solutions, visit **academic.cengage.com**

Purchase any of our products at your local college store or at our preferred online store **www.ichapters.com**

Printed in the United States of America
13 14 15 16 17 18 11 10 09

# Contents

# Preface to the Instructor

This book evolved from my years of teaching mathematics and working with students who had not done well on my tests. Many of these students did poorly despite the fact that they appeared to have been working hard. By analyzing the study behaviors of successful students, I accumulated many practical methods and suggestions that helped students improve their study skills and increase their level of achievement. This book is a result of that effort. I have tried to respond to almost any study difficulty a student might have in a mathematics course. I believe that no matter what problem a student has, there is always concrete advice you can give—advice more specific than just telling the student to work harder.

*Mastering Mathematics* tells students how to:

- develop an appropriate attitude
- deal with class notes
- approach homework

- ask good questions
- study more effectively for tests
- use supplemental resources
- avoid "mental blocks"
- respond when they think they are "lost"
- check their work

Users of the second edition will want to note some of the revisions made in the third edition. Many of the Personal Worksheets have been changed to provide a greater variety of activities through which students can determine their own individual study deficiencies and how to eliminate them. Chapter 5 from the second edition has been broken into two chapters in the third edition—Chapter 5 and Chapter 6. Chapters 9 and 10 from the second edition have been combined into one chapter, now Chapter 10. These organizational changes have been implemented to make all chapters more uniform in length.

A number of sections have been added in the third edition. Two of these are in the first chapter: one tells students how to find their own reasons to be dedicated to the course, and another describes how math anxiety can be reduced or eliminated by achieving success in the course. Chapter 4 includes a new section that provides specific suggestions for students in cooperative learning courses. In Chapter 6, there is now a section called "Complete Computer Homework Effectively," since more and more math classes involve computer work. Chapter 13 includes a new section, "Use a Math Tutor Successfully." In addition, "Practice" sections have been added to the ends of Chapters 5 through 13. These give students an immediate opportunity to apply some of the strategies described in the chapter.

If you are a mathematics teacher, *Mastering Mathematics* can enhance your effectiveness. By providing basic study skills to students who need them, it can free you to concentrate on teaching the subject matter.

If you are a study skills teacher, you will find this book unlike general study guides. It provides you with techniques that focus specifically on mathematics courses.

My experience suggests that by implementing more effective study techniques, all students are capable of improving their achievement in mathematics courses. The goal of this book is to help students realize this potential.

## Acknowledgments

Many people were generous with their time and gave useful and insightful feedback in the development of the original manuscript: Connie Cameron, Karan Collins, Lynn Boloyan, Jessica Doucette, Jennifer Kelly, Mary Mann, Alan Olinsky, Barbara Tavares, and Steve Wietrecki. I appreciate Martin Rosenzweig for his insights on the benefits of cooperative learning in mathematics. I'd also like to thank the following reviewers of the first, second, and third editions, who provided many valuable suggestions about the manuscript. The reviewers of the first edition were James Bolen, Tarrant County Community College; Loretta M. Braxton, Virginia State University; Susan Dimick, Spokane Community College; Charlotte K. Lewis, University of New Orleans; Gael Mericle, Mankato State University; Laura Moore-Mueller, Green River Community College; Patricia Pacitti, SUNY College at Oswego; Susan H. Talbot, Evergreen Valley College; John Waterson, Arkansas Technical University; and Gerry C. Vidrine, Lousiana State University. The reviewers of the second edition were Gabrielle Andries, University of Wisconsin, Milwaukee; Lois J. Dapsis, Quinebaug Valley Community College; Elizabeth J. Ince, United States Military Academy, West Point; Julie Keener, Central Oregon Community College; Thomas R. Love, Sul Ross State University; and Keith Worth, Scottsdale Community College. The reviewers of the third edition were Pamela Cohen, New Hampshire College; Janet Evert, Erie Community College; Kelly Ann Jackson, Camden County College; Julie

Keener, Central Oregon Community College; Kathleen Kraemer, Santa Rosa Junior College; Peter Moore, Northern Kentucky University; Chris Siragusa, Cypress College; and Shirley Thompson, North Lake College.

Marjorie Fuchs patiently read many drafts of the manuscript, contributing numerous insights and specific suggestions on the content, organization, and style. This book has benefited from the numerous discussions on pedagogy that I have had with her. In addition, she has been an inspiration to me throughout the project.

Thanks to Molly Heron and Tom Pyle for their creative ideas concerning the cartoon development and to Kevin Opstedal for his fine renditions.

I appreciate the careful work of the production team, whose members always responded to my suggestions with open-mindedness and good humor. I am grateful to Angie Gantner, my editor for the first edition, for her belief in the project from the beginning and for her enthusiasm, energy, and intelligence in directing its development.

Finally, special thanks to my students, past and present, whose aspirations and struggles inspired this project. This book is dedicated to them.

*Richard Manning Smith*

# Introduction to the Student

Have you ever been lost as to what systems, strategies, or methods you might use to achieve top grades in a math course? Have you ever thought you'd worked really hard in a math course but your grade just didn't show it? Have you ever had what you thought was a "mental block" during a math test and wished you could somehow eliminate your chance of ever having one again?

If you answered yes to any of the above questions, and if you are really determined to improve your results in your current or next math course, then this book is for you.

This book will help you to improve both your learning and your grades in any math course. You can do this even if you were never any good in math or have a weak background in math. You can do this even if you have always had mental blocks when taking math tests. You can do this even if the only grades you have ever received have been C's, D's, or F's. This book will tell you how you can turn those grades into A's.

There is a catch, though. The catch is that you may have to work harder. You certainly will have to work smarter!

## HOW THE BOOK IS STRUCTURED

The book begins with a diagnostic survey. This will help you to determine the areas in which you need to improve your approach to math courses, both in your attitude and in your study habits. Part I, "Make a Fresh Start," tells you how to begin the course with a positive attitude and gives you an overview of the course preparation system. Part II, "Make Successful Course Preparation Your Consistent Routine," presents strategies you can use even before the course starts that will help you learn more and achieve a high grade. Specific practical advice on how to handle class time, class notes, study time, and homework follows.

Part III, "Make Preparing for Tests a Sure Thing," is the most important part of the book. Here is where you will find specific strategies for preparing for math tests—strategies that will lead to increased confidence in your ability to do well on these tests.

The book contains numerous suggestions for activities that can help you succeed in a math course. You don't need to do *everything* I recommend. Choose at least some of the activities that you don't do now that you think might help you. If your present or most recent experience with a math course has been unsuccessful, you will surely have to do more than you have been doing. The diagnostic survey should give you good clues as to what kinds of improvements in your study habits might benefit you.

The chapters can be read in order and seen as providing an organized study system. The most important points are highlighted in boxes, while other significant points are highlighted in boldface type. Most chapters contain a summary list of questions and answers to help you remember the key points. The book can be a valuable reference for study ideas you can apply anytime you might want to improve your learning and your grades in a math course.

Finally, the book provides you with the opportunity to keep track of your own progress in improving your study habits in a math course. At the ends of all chapters are "Personal

Worksheets" that will provide you with an opportunity to pinpoint the strategies in the chapter that you think will benefit you the most. At the end of each Personal Worksheet in Chapters 5 through 13 is a "Practice" section. These sections encourage you to apply some of the strategies suggested in the chapter just after reading it, rather than waiting until you've finished the entire book. In addition, a "Study Habits Improvement Check" in Appendix A lets you evaluate the progress you have made in your study habits after reading the book and using the techniques.

This book is dedicated to helping you reach your goal: to make your next math course a major success. Such a success will have the additional benefit of helping you in future quantitative courses. For one thing, your attitude will be improved, since you will have proved that you can succeed. Furthermore, you will feel confident that you know how to use an effective system that leads to high achievement.

# How to Evaluate How Well You Study in Math Courses

In order to identify the areas in which you need to improve the most, you should complete the following survey. The scoring method is described at the end of each section.

## COMPLETING THE SURVEY

For each statement, circle the number that most closely indicates the degree of your agreement or disagreement.

### I. Your Attitude and Approach in Math Courses

| | *strongly disagree* | | | | *strongly agree* |
|---|---|---|---|---|---|
| 1. I usually believe that I can do well in math courses. | 1 | 2 | 3 | 4 | 5 |

| | *strongly disagree* | | | | *strongly agree* |
|---|---|---|---|---|---|
| 2. I am usually enthusiastic about learning in math courses. | 1 | 2 | 3 | 4 | 5 |
| 3. I believe I am good at math. | 1 | 2 | 3 | 4 | 5 |
| 4. I work persistently in a math course, regardless of how well I do on the tests. | 1 | 2 | 3 | 4 | 5 |
| 5. I usually enjoy taking math courses. | 1 | 2 | 3 | 4 | 5 |

## II. Classroom and Homework Habits

| | | | | | |
|---|---|---|---|---|---|
| 1. I miss at most two class hours per semester. | 1 | 2 | 3 | 4 | 5 |
| 2. I always get to class on time. | 1 | 2 | 3 | 4 | 5 |
| 3. I usually find it easy to understand what goes on in math classes. | 1 | 2 | 3 | 4 | 5 |
| 4. If I do not understand something in class, I will usually ask the teacher about it. | 1 | 2 | 3 | 4 | 5 |
| 5. I usually take clear and complete notes in a math class. | 1 | 2 | 3 | 4 | 5 |
| 6. I usually read my class notes carefully before the next class. | 1 | 2 | 3 | 4 | 5 |
| 7. I almost always make a persistent effort to do my homework before the next class. | 1 | 2 | 3 | 4 | 5 |
| 8. If I have questions arising from the homework, I ask my teacher or another student. | 1 | 2 | 3 | 4 | 5 |

| | *strongly disagree* | | | | *strongly agree* |
|---|---|---|---|---|---|
| 9. I find a way to check my solutions to homework problems before the next class. | 1 | 2 | 3 | 4 | 5 |
| 10. I frequently discuss homework and class notes with other students. | 1 | 2 | 3 | 4 | 5 |
| 11. If I have trouble understanding the textbook, I find other ways to master the concepts. | 1 | 2 | 3 | 4 | 5 |
| 12. Even if I understand most of what goes on in class, I am usually careful to do the homework before the next class. | 1 | 2 | 3 | 4 | 5 |

## III. Math Test Preparation Habits

| | | | | | |
|---|---|---|---|---|---|
| 1. I obtain or make a list of all the topics that may appear on the test. | 1 | 2 | 3 | 4 | 5 |
| 2. I write solutions to several problems on every topic, one at a time, before looking at the answers. | 1 | 2 | 3 | 4 | 5 |
| 3. I never leave most of my studying for the test until the day before the test. | 1 | 2 | 3 | 4 | 5 |
| 4. I work on one topic until I master it, and only then do I go on to the next topic. | 1 | 2 | 3 | 4 | 5 |
| 5. I make sure that I master every topic that might be on the test. | 1 | 2 | 3 | 4 | 5 |

| | *strongly disagree* | | | | *strongly agree* |
|---|---|---|---|---|---|
| 6. I can explain to another student how to solve all the types of problems that may appear on the test. | 1 | 2 | 3 | 4 | 5 |
| 7. I always study well enough not just to pass or to get in the 70s or 80s, but to get close to 100 percent. | 1 | 2 | 3 | 4 | 5 |
| 8. For each type of potential problem, I can describe the typical errors a student might make in solving such a problem. | 1 | 2 | 3 | 4 | 5 |
| 9. I study *all* possible topics that I might be tested on, even if I believe that the teacher is unlikely to include such topics on the test. | 1 | 2 | 3 | 4 | 5 |
| 10. I can identify the types of problems I am faced with even when the problems are placed in random order. | 1 | 2 | 3 | 4 | 5 |
| 11. Even though I attend class regularly, take complete notes, and do all the homework, I make an additional special effort to study for the test. | 1 | 2 | 3 | 4 | 5 |
| 12. I obtain a collection of problems and questions that can serve as a practice test. I write answers to all problems on the practice test without looking at the solutions. | 1 | 2 | 3 | 4 | 5 |
| 13. I usually know the material so well that I enjoy taking the test. | 1 | 2 | 3 | 4 | 5 |

## SCORING THE SURVEY

Score each question based on the number you circled. For each of the three parts of the survey, find your total number of points, and then refer to the appropriate point category below.

### I. Your Attitude and Approach in Math Courses

| | |
|---|---|
| 22 or above: | You approach a math course with a healthy attitude that tends to help you do well in the course. Your attitude and approach are not causing you problems. |
| 18 to 21: | Your attitude does not greatly handicap you in a math course, but you could definitely improve it. (See Chapter 1.) |
| 17 or below: | It is very likely that your attitude has been a major handicap to your doing well in a math course. (See Chapter 1.) |

### II. Classroom and Homework Habits

| | |
|---|---|
| 54 or above: | Your classwork and homework efforts are very good, and you do not need much improvement in this area. |
| 40 to 53: | Your classwork and homework preparation are about average. You can definitely make improvements in your preparation that will help you in the course. (See Chapters 2–6.) |
| 39 or below: | Your classwork and homework habits have been greatly handicapping you in math classes. You need to make significant improvements in your preparation that will help you in the course. (See Chapters 2–6.) |

### III. Math Test Preparation Habits

| | |
|---|---|
| 60 or above: | Your test preparation habits may be excellent. If you have circled any number less than 5, however, look at that statement again for clues on how to improve these habits. |
| 45 to 59: | Though you have no extreme test preparation problems, you can definitely improve your test preparation habits. (See Chapters 7–13.) |
| 44 or below: | Your test preparation in math courses can be greatly improved. (See Chapters 7–13.) It is likely that improvements in your classroom and homework habits may also help you to prepare for tests. (See Chapters 2–6.) |

# PART ONE

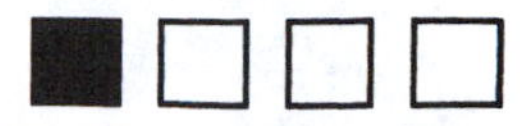

# Make a Fresh Start

# How to Have the Right Attitude in Math Courses

## BEGIN WITH AN OPEN MIND

The most important quality that will affect your success in a math course is your attitude. Your attitude determines what you will be willing to do in the course, and it is the quality of that effort that contributes most significantly to your success.

Take a look at these student comments:

"I've always hated math."

"I've never been good at math."

"I have mental blocks during math tests."

"Other students seem to do well in math with little effort."

"I have a weak math background."

"I haven't taken a math course for many years."

"I study for hours and still don't do as well as I want to."

These are some of the typical comments made by students who have a negative attitude toward math. Whatever your present attitude toward math, it has undoubtedly been influenced by your past experience in the subject. You may have been a poor math student for as long as you can remember. Perhaps only your most recent math experience has been a difficult one. You may think your background in math is deficient. Or it may have been a long time since you have taken math. For any of these reasons, you may have very little confidence in your ability to do well in math courses in the future. Similarly, if you have never enjoyed math before, you are not likely to look forward to taking such courses now.

## MOTIVATE YOURSELF TO BE DEDICATED

In order to be successful in your math course, it helps a lot if you can give yourself a reason or reasons that you want to be successful in the course. Here are some possible reasons for you to want to be successful in your math course:

- You need to learn the math as background for some other course you want or need to take.
- You need credit for the course to graduate.
- You need to achieve a certain grade in the course to get into college or graduate school.
- You need to learn the math as background for a job you want.
- You need to achieve a certain grade in the course to get a particular job.

If any of these reasons motivate you to want to do well in your math course, identify them now, and remind yourself of them from time to time.

**Find compelling reasons to dedicate yourself to your math course.**

Let me give you another reason for wanting to do well in your math course—one you may not have thought of. If, in the past, math has been very difficult (and perhaps unpleasant) for you, just imagine how you would feel if you could be truly successful for the first time. A positive math experience can provide a wonderful model for success, so that in the future, whenever you encounter some task, job, or project that appears at the start to be extremely difficult, you can inspire yourself by recalling your success in your "difficult" math course. This may be one of the best reasons of all for you to commit yourself to a successful math experience.

**If you can achieve success in a "difficult" math course, your awareness of that success can inspire you to pursue challenging projects in the future without becoming demoralized. This may be one of the most compelling reasons for you to strive for success in your math course.**

## OVERCOME YOUR "LOW ABILITY"

Low grades in the past do not necessarily indicate low natural ability. I have numerous examples from my teaching experience showing that some students who thought they had "low ability" in math began to be more successful than students who thought they had "high ability." This can happen when the "low-ability" student has a change of attitude, decides to take the course seriously, and plans to do the work necessary to be successful. In this same course, a "high-ability" student might become overconfident as a result of previous success and decide not to work very hard. Such lack of effort can result in underachievement for a "high-ability" student.

A striking example of what a "low-ability" student can do occurred a few years ago in a statistics course I was teaching.

After the first class, a student came up to warn me that she would need lots of help in the course. Jennifer said she was very worried about the course but was willing to work hard. I told her that if she needed help, she should not hesitate to ask for it. Since she was so concerned, I made a particular point of observing her progress in the course.

Jennifer sat in the front of the room in every class and paid close attention to what was going on. She obviously made significant attempts to do the homework, since she asked specific questions about it in almost every class. Jennifer was so careful not to fall behind or miss any point that came up in class or in the homework that she was able to help her classmates in the course at test time. She told me that they were smarter in math than she was but had neglected to do the homework on time.

The result of all her effort was that Jennifer had the highest average of all my statistics students *for the entire year*, receiving at least a 97 on every test! It got to the point in the course where I thought that she had been joking on that first day about how difficult she thought the course was going to be for her. She insisted that she had not been joking and that it was only her exceptional effort that was responsible for her success.

The success of a "low-ability" student with a constructive attitude and the lack of success of a "high-ability" student with a poor attitude blurs the distinction between "low-ability" and "high-ability" as labels for math students. After all, if it is possible for some "low-ability" students to be more successful than "high-ability" students, which students actually have the higher ability? Of course, there are some real differences in natural ability among students. However, the significance of these differences can be lessened and, in some cases, overcome.

The bar graph on the next page shows how a student's natural ability does not always relate to the student's achievement in a course. The bars suggest that a student's achievement can be seen as some combination of the four resources: natural ability, previous knowledge of the content, total study time, and use of study time. Student C's achievement results from an equal emphasis on all four resources. Student B's achievement

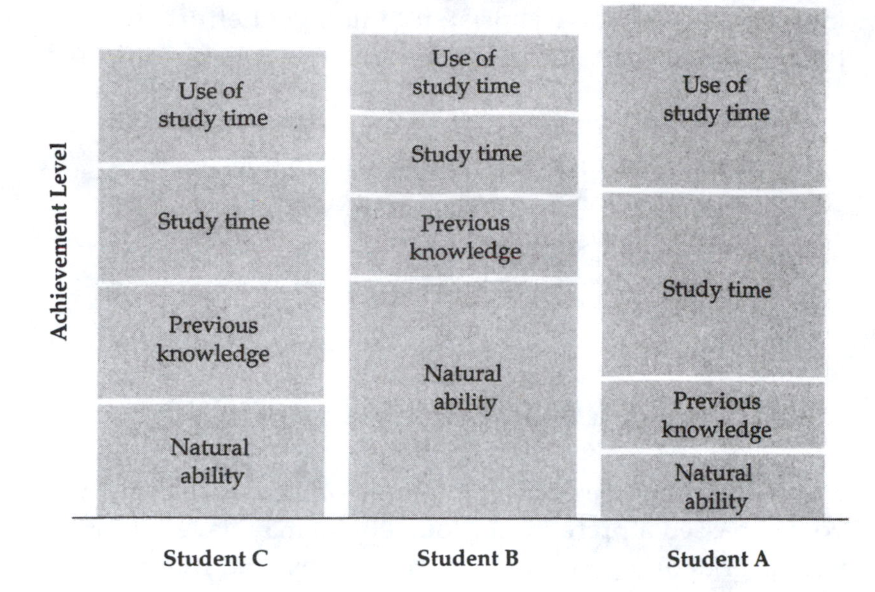


is greater than that of Student C, mainly because of student B's higher natural ability. This achievement is greater, even though the student makes less effective use of study time than does Student C. Student A, with the lowest natural ability and the lowest previous knowledge of the three students, produces the greatest achievement because of a greater willingness to devote study time (dedication and persistence) to the course, and because of more effective use of that study time.

You are not in control of whatever your natural ability is, but you *are* in control of how much time you devote to the course (dedication and persistence) and how you use that time. The purpose of this book is to help increase your achievement in math courses by inspiring you to be more dedicated to the course and, most significantly, by showing you how to make the most effective use of the time you use. This should be good news for you if you have had unhappy experiences in previous math courses but have now decided to make a determined effort to be successful. Regardless of your poor past performance, your chances of success are greatly increased if you make consistent, intelligent efforts toward pursuing that success.

> **Overcome your "low ability" in math courses with attention to detail and a persistent, "never-quit" attitude.**

## OVERCOME A WEAK MATH BACKGROUND

Many math courses require taking prerequisites. Even if you've passed a prerequisite course, you may have forgotten the material or may never have really mastered it in the first place.

A typical example occurs when a student who is required to take a calculus course in college has a deficient algebra background from high school. What should you do if you are in a position like this? If your background is exceptionally poor, the best approach is to register for an algebra course in the term *preceding* that of the calculus course you are planning to take.

Whenever you take a remedial course, follow these steps to derive the greatest benefit:

1. Begin with a positive attitude. Decide that you will take the course seriously whether the course is for credit or not.
2. Find a good teacher for the remedial course.
3. Obtain a copy of the textbook and any review books.
4. Do not miss a class or arrive late.
5. Do the homework regularly.
6. Study for and take all tests.

In general, apply the methods described later in this book to the remedial course before you begin the course that follows it.

**If you have an extremely poor background for the math course you are planning to take, first take a remedial course even though you may not receive credit for it.**

What if your background is only moderately weak?

**If you have only a moderately weak background, do not ignore it. Patch up your weaknesses in the course as soon as you encounter them.**

If you decide that this approach can satisfactorily improve your background, follow these steps:

1. Obtain for your reference during the course one or more review books or textbooks dealing with the subject in which you are weak. Make sure these books present the material at a level you can understand.
2. If you have trouble understanding a topic in the course because of your background deficiency:
    a. Identify the deficiency as soon as possible and as specifically as you can.
    b. Find the section of one of the reference books you have found that deals with your deficiency.
    c. Read the section and practice working out problems from that section.
    d. Ask your teacher questions.
    e. Get a tutor based on a recommendation from your teacher or other students.

When you have "patched up" your deficiency, return to study the topic you did not understand in your current course. You should now find it easier to deal with.

## OVERCOME A BIG TIME GAP BETWEEN MATH COURSES

If you haven't taken a math course for a few years or even for many years, you may be overestimating how much the time gap will handicap you in your next math course.

I have known many students who had small or large time gaps since their last math course. In all of the cases when the student had sufficient background for the course, the handicap evaporated after a few weeks in the new math course. What happens is that, usually, the student's comprehension level in the new course returns to whatever level it was in the last math course, no matter how long ago that course was taken.

One problem you might have can arise if your new course requires a certain level of knowledge of computers or calculators. (This requirement may not even be stated in the list of prerequisites.) Since these tools may not have been used in your last math course, you may lack this knowledge. For this reason, if a large amount of time has passed since your last math course, speak to the teacher before the course starts to determine what knowledge of the computer, the calculator, or any other subject not listed as a prerequisite is required to be successful in the course.

If you were successful in your last math course, the time gap will not keep you from being successful in the current course. If you were not successful in your last math course, however, you will not be satisfied to return to your old level in the current course. You will want to do much better.

One way to do that is to follow the suggestions in the last section on overcoming a weak background. In summary:

**If you have a large time gap since your last math course, you might take a remedial course before taking the current one, even though you may not receive credit for it.**

**If you have only a small time gap since your last math course, patch up your weaknesses in the course immediately as you encounter them.**

## OVERCOME MATH ANXIETY

Some students blame their lack of success in math courses on their "math anxiety." If their theory is correct, what limits their ability to achieve in future courses is their history of tension, stress, and lack of success during previous math classes or math tests. There are entire books whose goal is to help people overcome their math anxiety by using a psychological approach. However, reducing math anxiety (as nice as that may feel) does not automatically lead to success. You *still* need to use effective study techniques. Therefore, your focus should not be on lowering math anxiety in order to increase success. Achieving success will lower math anxiety. The first goal of this book is to provide specific, concrete methods for helping students become successful in their math courses. Any lowered math anxiety will be a by-product of that success.

**Lower your math anxiety by achieving success in math courses using specific, effective methods.**

## DON'T MAKE EXCUSES

Sometimes, students overcommit themselves by registering for more courses than they have time for. If your job requires you to work forty hours per week, and you also intend to register for five challenging courses, you might want to reconsider your plans. In particular, do not even sign up for a math course unless you know you will be able to devote a good deal of time to it.

Once you are in a math course and know that you need to complete it, ask yourself frequently what you need to do to be successful in the course. Forget about why you haven't been successful in the past and why you can't be successful in the future. Do not use these as excuses for not doing well now.

Take responsibility for your own education. If you are committed to staying in the course, make up your mind that you will not blame your teacher, your background, your past performance, your personal deficiencies, or difficulties in your life for problems you may encounter in the course. Make up your mind that you will do whatever it takes to master the course *without making excuses for your lack of success.* (This book will suggest what you can do.) There is nothing a teacher can do that will make up for what a student won't do.

What about students who have a learning disability? Don't they have a legitimate excuse for not doing well? Some students do have a learning disability that has been diagnosed by a guidance counselor, psychologist, or psychiatrist. These students and their parents sometimes mention the learning disability as the reason the student has not done well in previous math courses. Even if you have such a learning disability, applying the techniques in this book should still improve your achievement in future math courses. What is most important is that you don't use the existence of the learning disability as an excuse for not making a dedicated and persistent effort in future math courses.

The effects of all of your "handicaps" in the course can be either decreased or overcome. You can always think of excuses for not doing well. Assuming you have sufficient preparation for your current math class, you need to:

**Persist in the course regardless of any problems that might arise.**

What might happen if you become significantly more successful in math courses? Maybe you will begin to enjoy the subject, since it may have been your lack of success that made you dislike the subject in the first place!

## TAKE A POSITIVE APPROACH TOWARD THE COURSE

Regardless of any negative experiences you have had in the past, you must make four decisions *now* about how you will approach this course:

**To have a positive approach:**

1. **Recognize that you have control over how well you will do in the course.**
2. **Decide now that you will make an honest effort to do well in the course.**
3. **Decide now that you will work not merely to pass the course but to do much better than pass.**
4. **Decide now that you will persist in working hard in the course until the end, *regardless of any setbacks that might occur along the way.***

It should make you feel better to know that any difficulties you may have had in past math courses can be greatly reduced if you have a system for success. This book will give you such a system.

This system will show you how to be successful in spite of any previous negative experience you may have had in math, *and* in spite of any feeling or perception that you lack mathematical ability.

## MAKE THE RIGHT STATEMENTS TO YOURSELF

Research has suggested that in order to maintain the most helpful attitude at the beginning of a course, you should make appropriate positive statements aloud to yourself. For example, say to yourself with enthusiasm:

> "I have control of how I will do in this course."
>
> "I am going to enjoy taking this math course."
>
> "I will always make an enthusiastic effort to do well in this course."

"I will persist in working hard in the course until the end, *regardless of any setbacks that might occur along the way.*"

"I am determined to use all my resources to get an A in this course."

If you have any difficulty believing any of these statements about yourself, that's all right. Say them aloud and act *as if* you believe them. Continue to behave this way throughout your course. For example, if you don't really believe that you will enjoy the course, act *as if* you will enjoy it, and make the efforts you would make if you did enjoy it. As you immerse yourself in the course with greater effort and thought than you are used to making, you will start to enjoy the feeling of success that comes from understanding more than you usually do. When you begin to do better on tests, you may naturally start to enjoy the course. At this point, you may find it even easier to prepare for future tests.

**Act as if you have control of your level of success in a math course. Act as if you are really enjoying it. Eventually, your habit of pretending and your resulting success will make your feelings match your behavior.**

## MAKE AN EXCEPTIONAL EFFORT FROM THE BEGINNING

Many students begin a course by being overly relaxed. After just returning from vacation, they choose to take a leisurely approach to beginning their coursework. They believe that such an approach is acceptable, since they probably will not have to worry about a test for several weeks.

Taking this approach in a math course can be fatal. The cumulative nature of math courses can lead students to pay a very high price for procrastination.

I recommend that you take the exact opposite approach. The most effective way both to improve your attitude and to achieve the best results in the course is to:

**Be "overdedicated" for the first two to three weeks of the course.**

During this time, pay especially close attention to classwork. Attend all classes. Get there on time. Spend a great deal of time on homework. Be willing to ask questions about notes and homework in any class. Be overdedicated for these two to three weeks and you will most likely receive these benefits:

1. Your overdedication will lead you to learn more of the material at a faster rate than you have in the past.
2. Your increased understanding will improve your attitude toward the course and, perhaps, even to math in general.
3. Your exceptional effort will become a habit that you will find easy to maintain throughout the course.

## SUMMARY

### How to Have the Right Attitude in Math Courses

Try to answer the question on the left before you look at the answer on the right.

| | |
|---|---|
| 1. What is the most significant controllable quality that influences my success in a math course? | It is my attitude in the course. |

| | |
|---|---|
| 2. How can I become motivated to want to do well in a math course? | I should find compelling reasons to dedicate myself to the course. |
| 3. How can success in a math course benefit me in the future (even if I don't need math for my career)? | It could inspire me to pursue challenging projects in the future. |
| 4. How can I overcome poor past performance in math courses? | I can overcome poor past performance with attention to detail and a persistent, "never-quit" attitude. |
| 5. How can I overcome a poor math background or a large time gap since my last math course? | I can overcome a poor math background or a large time gap either by taking a remedial course or by patching up my weaknesses during the course. |
| 6. How can I lessen my math anxiety? | I can lessen my math anxiety by achieving success in math courses through the use of specific, effective methods. |
| 7. How should I respond to any setbacks? | I should persist in working hard in the course until the end, regardless of any difficulties or setbacks that might occur along the way. |
| 8. What should I do if I don't feel that I have control over my level of success in the course or am not enjoying it? | I should act as if I have control over my level of success in the course, and I should act as if I am enjoying it. Eventually, my habit of pretending and my resulting success will cause my feelings to match my behavior. |
| 9. What strategy should I pursue for the first few weeks of a math course? | I should be overdedicated for the first two to three weeks in the course. |

## PERSONAL WORKSHEET

I. Check the statements that are true for you.

___ 1. I have always hated math.

___ 2. I never expect to do well in a math course.

___ 3. I have "low ability" in math courses.

___ 4. I don't work hard at the beginning of a math course.

___ 5. I haven't taken a math course in years.

___ 6. I usually try just to pass a math course.

___ 7. I stop working very hard after I do badly on a math test.

___ 8. I have a very weak math background for my current or next math course.

**Other**: ________________________________________

________________________________________________

________________________________________________

II. Check the top five changes that you plan to make in your attitude and approach to math courses. Add any other strategies you plan to use.

___ 1. I will find reasons to dedicate myself to the course.

___ 2. I will have a persistent, "never-quit" attitude.

___ 3. I will take a remedial course first.

___ 4. I will patch up my weaknesses in the course as soon as I encounter them.

___ 5. I will act as if I have control of my level of success in my current (or next) math course.

___ 6. I will work not just to pass but to learn enough to obtain the highest possible grade in my math course.

___ 7. I will persist in working hard in the course until the end, regardless of any setbacks that might occur along the way.

___ 8. I will make an exceptional effort right from the beginning of the course.

___ 9. I will ask the teacher more questions than I have in the past.

**Other:** *I will* ______________________________

______________________________________________

______________________________________________

III. After you try these strategies, refer back to this page and highlight the ones that worked best for you.

# Make Successful Course Preparation Your Consistent Routine

# How to Begin *Before* Your Math Course Starts

## TAKE AN ACTIVE ROLE IN COURSE PREPARATION

Most students begin their work in a math course *after* they attend the first class and find out from the teacher what will be covered in the course. You can do much better than that if you are willing to take a more active and dedicated role before your course begins.

This chapter will give you some ways to get a big head start in any math course.

## SELECT YOUR TEACHER WITH CARE

In many school situations, you have no control over who your math teacher will be. Under these circumstances, you may be stuck with a teacher that you don't like. Chapter 12 will provide useful strategies to help you handle this situation. For now, however, suppose that there is more than one section of your math course available, and that you have a choice of which section to enroll in. If you can, several weeks before registration, obtain the names of all the teachers who are teaching the course.

Take your list around to friends and other students and ask them for information about the teachers on your list. Make sure you get several independent opinions about each teacher before you decide on the one you will select.

If you do have a choice at your school, on what basis should you pick your math teacher?

**Select a teacher who can or will:**

1. **Explain concepts clearly.**
2. **Welcome questions in class.**
3. **Help students outside of class.**
4. **Cover the entire syllabus.**
5. **Give fair tests.**
6. **Provide helpful handouts to complement the notes.**

Don't select a teacher who has the reputation for being an easy grader but who may not be a good teacher. If you do this, you will be at a disadvantage should you ever need to use the math from that course in later courses or in your career.

In selecting your teacher, you should also consider your learning style. Do you learn better alone or in a group? Do you prefer to listen to a lecture, or do you prefer a more interactive approach? Do you like a teacher who writes a lot on the board? Find out about a teacher's teaching style and try to choose one whose style matches your learning style.

What do you do if the teacher you select is so popular that his or her classes are filled before you are able to register? If it is possible at your school, go to the teacher weeks before the course starts and ask to be on a waiting list to get into the class. Tell the teacher that it is very important for you to get into the class. Emphasize that your interest is based on that teacher's reputation and that you will work very hard in the course. Be both polite and persistent.

If none of this is possible at your school, there are other things you can do.

## BUY THE TEXTBOOK EARLY

Once you know who your math teacher will be, you can determine from the teacher or school bookstore what textbook will be used in the course. If possible:

> **Buy your textbook early. Buy a new textbook or a used one that is not filled with markings from a previous student. Get a head start by reading appropriate sections before the course starts.**

It is important to have an unmarked book, since you will want to make your own markings during the course without any distractions. After all, you don't know what kind of student made the markings in the book before you bought it.

Determine from the teacher which chapters and sections will be covered during the first three weeks of the course.

Begin reading the appropriate sections during the two or three weeks before the course starts. Try to work out problems at the end of each section.

If you find that your assigned textbook is not providing you with sufficient help:

> **Before the term begins, go to the school library or local bookstore to find other relevant textbooks, review books, or study guides that can help you master concepts throughout the course.**

For example, go to the section of the school library that contains books dealing with the subject of your course. Examine those books that seem to be competitors of your assigned book. Look up several topics that you are having difficulty with in your assigned textbook. You can sign out any library books that you find useful and use them to enhance your understanding of the concepts. It is also helpful to go to your local bookstore to find any review books or study guides available for the course you will be taking.

You may find that studying the material early is a more pleasant experience than you would have expected. This activity provides you with the rare experience of studying a subject without having any immediate outside pressure. Without such pressure you may even discover that math can be fun! At worst, you are guaranteeing that you will be ahead of the class right from the beginning. This will increase both your morale and your understanding of the material.

Remember, though, that this experience should be a *positive* one. If you find that studying material before the course starts is confusing for you, stop doing it and wait for the course to begin.

## AUDIT THE COURSE FIRST

Some schools will allow students to sit in on (audit) a course before they actually take it for credit. If it is allowed at your school, this may be a good idea under some circumstances.

> **If you are planning to take what you think is an exceptionally difficult math course, you may wish to sit in on (audit) the course the term before you register for it.**

Most students do not have enough time to take this advice, and you don't *have to* do it to succeed in the course. However, if you do have the time, it can give you an extra advantage when you finally take the course for credit.

If you do decide to audit, begin by finding a teacher whose course you are considering enrolling in. Ask that teacher if you can sit in on his or her class without being officially registered for it. Many teachers welcome students to their classroom who demonstrate that much enthusiasm.

You should *not* audit the course unless you plan on treating the course as if you were actually registered for it. This approach requires that you take notes, do homework, ask the teacher questions, and study for and take all tests. The benefit of auditing this difficult course is that you can do all of this without the pressure of having to be graded. Then, when you actually take the course for credit, you will be able to succeed more easily.

## SUMMARY

### How to Begin *Before* Your Math Course Starts

Try to answer the question on the left before you look at the answer on the right.

1. How should I select my math teacher?

   Select a teacher who has a good reputation for teaching concepts clearly, for being willing to answer questions and offer outside help, for covering the entire syllabus, for giving fair tests, and for providing helpful handouts to complement the notes.

2. When should I buy my textbook and how should it look?

   I should buy my textbook early—either a new textbook or a used one that is not filled with markings from a previous student.

| | |
|---|---|
| 3. What about using additional textbooks and study guides? | I should find additional text-books and study guides before the term begins to increase my ability to master concepts throughout the course. |
| 4. Should I ever audit a math course? | To gain a big advantage, I could audit a math course before registering for it. |

## PERSONAL WORKSHEET

After reading this chapter, I plan to make the following three improvements in preparing for a math course before it starts.

*I will* ___________________________________________

# CHAPTER 3

# How to Master the Course Using Four Major Steps

## IDENTIFY THE FOUR KEY STEPS IN COURSE PREPARATION

What will it take for you to be successful in a math course? The system that will provide the tools for your success requires you to do several things:

> **In course preparation, take four key steps:**
>
> **1. Try your best to "see" or "follow" what the teacher is explaining in the classroom.**
>
> **2. Master both the class notes and the assigned textbook material.**
>
> **3. Solve homework problems and ask questions that deal with course material.**
>
> **4. Follow an organized study plan *specifically* for the test.**

Completing these steps will assure the most learning and the most success on tests in the course.

## FOLLOW THE TEACHER'S EXPLANATIONS IN CLASS

If you don't generally follow your teacher's explanations in class (step 1), you are likely to have difficulty reading your notes and textbook later (step 2), doing the homework (step 3), and following an organized study plan specifically for the test (step 4). You also are unlikely to do well on the test that follows. This may not be true if your textbook is much clearer than your teacher or if the test is unexpectedly easy. In most cases, however, following a teacher's explanations in class is a major help in achieving success on math tests. If you have trouble following your teacher in class, you can compensate for this difficulty by working extra hard at steps 2, 3, and 4.

## MASTER CLASS NOTES AND TEXT MATERIAL

Even if you follow your teacher's explanations in class, you may have trouble understanding your notes and textbook (step 2) once you are outside the classroom. This may make it more difficult for you to do homework (step 3). If this occurs frequently, then you are likely to have trouble with tests.

## WORK OUT HOMEWORK PROBLEMS ON TIME

What can happen to a student who tries to skip one or more of the above steps? Consider the case of Scott, a student in a finite math course. Since he is always able to follow his teacher's explanations and examples in class and understand his notes and textbook, Scott feels confident enough to neglect doing his homework until after several classes have passed. When he finally does the homework, he has great difficulty working out the problems. By test time, although he has followed his teacher's explanations and understood his notes and textbook, he hasn't derived the deeper understanding that would have resulted from keeping up with the homework.

Scott studies for the test, takes the test, and is shocked to receive a grade of 58. He says to his teacher, "I was able to understand everything you did in class. How come I got only a 58 on the test?"

Scott's problem is that he paid serious attention only to step 1 of the required *four* steps in successful course preparation.

## FOLLOW A SEPARATE ORGANIZED STUDY PLAN FOR THE TEST

You may assume that if you can "follow" your teacher in class (step 1), master class notes and textbook material (step 2), and you can solve the homework problems (step 3), you automatically ought to be able to do well on the test without special preparation. Often, however, this will not be the case.

In addition to listening closely to her teacher in class and understanding explanations and examples, Melissa, another student in Scott's class, makes sure to work out the homework problems after each class. If she has questions based on her homework efforts, she makes sure to ask them during the next class. As a result of her classroom understanding and diligent attention to homework, Melissa is so confident that when she studies for the test she works out only a few problems relating to some of the topics that may appear on the test. She takes the test and knows that she is not doing well since there are several problems she cannot do at all, and others she can do only partially. When her test is returned she gets the grade of 68. After her diligent work before the test, Melissa is depressed by her performance.

Melissa's problem is that she dedicated herself to steps 1, 2, and 3 but was negligent in completing step 4. She did not have a separate organized study plan specifically for the test.

To master a math course, you must seriously follow all four of the key steps to successful course preparation. The next three parts of the book will tell you, in detail, how to be successful in each of the four steps. This is the core of the successful course preparation system.

Specifically, step 3 requires greater mastery of the concepts than do steps 1 and 2, and step 4 requires greater mastery than does step 3. Therefore, Part II of the book describes how you can derive the greatest benefit from your classroom experience and from doing homework. Only after getting the most from class and homework can you benefit from using specific methods that can help you when preparing especially for tests. These methods are described in detail in Part III.

## SUMMARY

### How to Master the Course Using Four Major Steps

Try to answer the question on the left before you look at the answer on the right.

1. What four activities together will guarantee my success in a math course?

1. Trying my best to "see" or "follow" what the teacher is explaining in the classroom.
2. Mastering the class notes and the assigned textbook material.
3. Solving homework problems and asking questions that deal with course material.
4. Following an organized study plan specifically for the test.

2. What happens if I skip one or more of the four steps?

If I omit one or more of the four steps, I place myself at great risk of not doing as well on the test as I would like.

## PERSONAL WORKSHEET

I. Check the statements that are true for you.

___ 1. I have not been sufficiently dedicated to trying to understand math concepts as they are presented in class.

___ 2. I have not paid sufficient attention to mastering class notes and textbook material.

___ 3. I have not worked on homework in a sufficiently dedicated and timely manner.

___ 4. I have not followed an organized study plan specifically for the test.

II. Check the two most important changes that you plan to make in current and future math courses in completing the four steps in course preparation.

___ 1. I will be more dedicated to trying to understand math concepts as they are presented in class.

___ 2. I will be more dedicated to mastering class notes and textbook material than I have been in past math courses.

___ 3. I will be more dedicated to working on homework than I have been in past math courses.

___ 4. I will follow an organized study plan specifically for the test.

III. Rank the strategies you checked in section II according to how important you think they are for your success. Put the number 1 to the left of the most important, the number 2 to the left of the second most important, and so on.

IV. After you try these strategies, refer back to this page and highlight the ones that worked best for you.

# CHAPTER 4

# How to Use Class Time Effectively

*Note: If you scored very high on questions 1–5 of the "Classroom and Homework Habits" section of the diagnostic survey, you might choose to skip this chapter.*

## LEARN AS MUCH AS YOU CAN DURING CLASS TIME

The best way to begin successful course preparation is to persist in learning as much as possible *during the class time itself* and to leave the class with the clearest possible reference material (class notes and, perhaps, a cassette recording). The more diligently you do this, the easier it will be for you to do the homework and keep up with the pace of the course.

One way to learn more in class is to increase your participation. For example, if the teacher directs a question to the class, try to give the teacher an answer. Even if you are wrong, the attempt and the immediate feedback from the teacher will help you to increase your knowledge. In addition, your own involvement in the class will likely make the period more enjoyable for you.

When some students do not follow what the teacher is doing, they might say to themselves, "I can't understand this. I'll try to figure it out when I get out of class, so I won't bother to ask any questions now." What's wrong with waiting until later to try to clear up your questions?

If you wait, you are passing up the chance to obtain the answers directly from the teacher or other students at the time the class is dealing with the issue. You may find it difficult to get the answer later; in fact, you may even forget that you had the question! It is better to "strike while the iron is hot." The teacher will not be there when you get home.

Furthermore, when you clear up your confusion in class, you are saving your own time—the time you would have had to put in outside class to get your question answered.

**Feel free to ask questions in class.**
**Don't put off questions until later.**

You surely will not *master* a course just by paying attention and asking questions in class, but you will make mastering it a lot easier if you do.

There are some circumstances, however, when you may be better off *not* asking a question in class. Suppose you already have asked a few questions in a class period and your teacher has stopped encouraging questions from the class. If you have a question at this point, it might be more considerate of the class not to ask it now.

There are some situations, such as large lecture classes, where questions from students may not be allowed or are, at best, strongly discouraged. In these classes, you may have to put off your question until later. Nevertheless, do not ignore your question!

If you decide for any reason not to ask your question in class, write it down in the appropriate place in your notes or textbook. If your question involves confusing material in your notes or textbook, mark the appropriate lines with a question mark. This will make it easier for you to locate the lines later when you try to solve the difficulty in any of several ways.

**After class, discuss any question with your teacher or a classmate, or try to answer it yourself with the help of your notes or textbook or any other textbooks or study guides you can find. Persist until you get a satisfactory answer.**

Most students, when they ask a question in class, pay close attention to the answer they get from the teacher, but they don't listen as carefully when the teacher is answering *other students' questions*. For example, I have had students ask me how to do a problem from the book just before I passed out a quiz containing that identical problem. The student who asks the question always gets it right; however, other students often get it wrong, thereby demonstrating that they were not paying much attention to the student's question.

**It can be very helpful for you to listen carefully to other students' questions and to the answers the teacher gives them.**

## ATTEND ALL CLASSES

Many students grossly underestimate the harm they do to themselves when they miss even one class.

**Attend all classes. Missing even one class can put you behind in the course by at least two classes.**

My experience provides irrefutable evidence that you are much more likely to be successful if you attend *all* classes, always arriving on time or a little early. If you are even a few minutes late, or if you miss a class entirely, you risk feeling lost in class when you finally do get there. Not only have you wasted the class time you missed, but you may spend most or all of the first class after your absence trying to catch up.

Do you think that it doesn't matter if you miss your math class once in a while, or if you show up a few minutes late to a class? Maybe you think that it is silly to make a big deal about such a trivial issue. Consider, however, the following scenario.

Suppose you miss one of your math classes. Then, between that class and the next class, you do not bother to contact another student to learn what material was covered and what homework was assigned.

Since you do not want to miss too many classes in a row, you do attend the next class. At the beginning of that class, some students are asking questions about the homework and the notes from the last class. Since you did not get the notes or do the homework, the students' questions and the teacher's answers do not make much sense to you and you feel lost.

After the questions are fully dealt with, the teacher begins to teach new material. While presenting this material, the teacher assumes you have read all the previous notes and have completed all the assigned homework. Clearly, the teacher expects you to have at least a modest understanding of the most recently covered material. Since you have neither done the homework nor mastered the notes, the teacher is moving too fast for you. In short, as a result of your negligence, you risk having a more difficult time understanding the new material than do students who have kept up with the work.

By the end of the class, you have understood very little. You will now be behind, not just one class, but two classes. Since you did not follow much of the material in this new class, you face the prospect of having trouble both following these notes and doing the new homework. At this point, you will have developed a major morale problem.

Let's say that after this second class you decide that you want to catch up in your math course. At this point, you need to:

1. Obtain a clear set of notes and the homework assignment from the previous class.
2. Read and understand that set of notes.
3. Work out the homework problems from the previous class.
4. Read and understand the set of notes from the class you attended.
5. Work out the new homework for that class.

---

All of these tasks need to be accomplished before the next class or you will fall even more behind. Furthermore, understanding the notes from the class you missed is likely to be a lot more difficult and time-consuming, because you will be attempting to learn the material without the help of your teacher. Finally, that one missed class makes the task of preparing for your next test in the course that much more challenging. You don't need extra challenges like this!

You make yourself extremely vulnerable to the above scenario if you miss *just one* math class and don't obtain notes or do homework before the next class. Let's compare the experiences of two students.

Jamal missed a class in his statistics course and neglected to obtain the notes or do homework to prepare for the next class. When he went to the next class, he found it difficult to follow the teacher's explanations and notes. He didn't even understand enough to ask any questions.

When Kristen missed one of her statistics classes, later that same day she called a student in her class who gave her the homework assignment for the next class and allowed her to copy the class notes. Kristen then made a dedicated effort to study the notes and work out the homework problems before the next class. She prepared so well that when she arrived at that class she had no trouble understanding the teacher. She was able to understand other students' questions and even asked one of her own. Kristen's efforts at obtaining the notes, studying them, and then doing the homework helped her to overcome the potential harm that might have resulted from missing that one class.

Suppose you do not intend to miss any of your math classes, but because of illness or some other significant reason you must miss one. Then what should you do?

What you need to do is to follow Kristen's example when she missed her statistics class.

**If you miss a class:**

1. **Get an accurate copy of the notes from the class that you missed in addition to the homework assignment and any handouts that the teacher may have provided.**
2. **Read the notes and the corresponding sections of the textbook and make a serious attempt to do the homework.**

When you get to the next class, you should be as prepared as possible. Ask some questions in this class. Be involved.

In addition to attending all classes, you need to:

**Arrive on time or a little early, get out your notes and homework, and identify any questions or comments that you plan to bring to your teacher's attention once the class begins.**

If you think that it does not matter very much if you get to the class, say, ten minutes late, you are wrong. If you are ten minutes late, you may miss a discussion of solutions to homework problems that you had questions about, or you may find yourself several minutes behind in taking notes. In addition, arriving late is unlikely to win you any popularity points with your teacher!

If you are a few minutes behind in the notes, you put yourself in the position of having to do too many things simultaneously. You have to copy and try to understand the notes you missed at the same time that the teacher is putting new notes on the board. In this situation, you are likely to fall even further behind than you were when you walked into the classroom.

Any extra moments before class can be used to review the notes from the previous class and the homework assignment. Try to think of any new questions for the teacher that did not occur to you before.

## ATTEND MORE THAN ONE SECTION OF THE COURSE

If it is possible at your school and you have the time:

**Attend more than one section of the course.**

At some schools, an instructor may teach more than one section of the same course in the same term. If you are enrolled in one of these sections and you like your teacher, you might want to sit in on the teacher's other section as well. Most teachers will let you to this, and if you have the time, it can really help you deal with your notes. For example, in the first of the two classes, you can focus on trying to write a complete set of notes and understand as many concepts as you can. In the second class, you get a repeat exposure to the concepts without the burden of taking notes. As a result, you can pay more careful attention to what the teacher is saying and identify any questions that you might have.

You can also try to attend sections of the same course taught by another teacher. This can give you the advantage of seeing different presentations of the same material.

## SIT IN THE FRONT OF THE CLASSROOM

> **The best place to sit in a classroom in a math class is as close to the front and center of the room as you can.**

There are several reasons for this piece of advice. First, when you sit in the front and center part of the classroom, you can clearly see both the teacher and all sections of the board. In addition, there will be few or no students in your line of vision. This is especially important in a math class, since math

teachers regularly rely on visual aids such as the board, overhead projectors, or computer monitors, to explain or illustrate their points.

In addition to being able to see better, you can also hear the teacher better. Sitting up front helps you pay attention to teachers who speak softly or who have their backs to the class as they write on the board.

For these reasons, if you sit "front and center," you will find yourself more involved in what's going on in the class. This might influence you to ask questions when you need clarification, and respond when the teacher asks questions of the class. Being in front makes it easier for the teacher to hear you when you ask a question. Furthermore, if you tend to be shy or inhibited, it may be easier for you to ask questions when you sit in the front of the room, since you don't have to experience a large number of students turning around to look at you.

## ORGANIZE YOUR NOTEBOOK

**Use one large loose-leaf or spiral notebook devoted exclusively to math. Use the first half for class notes, the second half for homework.**

It is important that you have two sections in your notebook, one for classwork and one for homework, so that when you refer to a notebook problem you know whether it was one the teacher presented in class or one that you did for homework.

Begin each day's notes by placing the date at the top to make it easier to find the specific sections you need when you want them. Make sure to include any topic or chapter heading that your teacher provides to help you identify notes.

Some students prefer loose-leaf over spiral notebooks. Loose-leaf notebooks provide a great deal of flexibility. They allow you to add a page of notes at any time, or to rewrite notes to improve their clarity. When you are doing homework, a loose-leaf notebook also encourages you to attempt and to write down different approaches to problems. When you finally have a correct solution, you can insert a neat copy of it into your notebook and remove any unnecessary pages. If you use a loose-leaf notebook, apply reinforcements to all of the holes of the pages before the course starts.

Regardless of the kind of notebook you prefer, make sure to leave room for additional notes. One way is to leave a two-inch margin on the left; another is to use only one side of each page for your first round of note-taking. You may also choose to do both. These strategies will allow you later to add supplemental notes, clarifications, and examples. The specific ways that you can add such notes will be discussed in Chapter 5.

## TAKE COMPLETE CLASS NOTES

What should you include in your math notes? Some students try to write out only the main ideas, and some books actually recommend that you do this. However, this is the wrong thing to do. Copy all notes written on the board. If you take down only the "main ideas," you will have a much more difficult time understanding your notes later.

However, even copying all the notes on the board will not give you a complete set of notes. First, many teachers supplement their board work with helpful explanations that they say but do not write down. Second, sometimes a student makes a comment that may be helpful to your understanding. In each of these situations, make sure you add the new explanations or insights to your notes. The more complete your notes, the better you will understand them later.

Some math teachers provide a very complete set of notes on the board. They give many illustrations, and when they work out an example or write out a proof, they provide a comprehensive list of steps, making it easy for most students to connect each step to the next. When the board notes are written this completely, you may not need to ask many questions in class.

**Take a complete set of notes in class. Add any helpful clarifications to your notes that you hear in class.**

## ASK QUESTIONS ABOUT CLASS NOTES

What should you do when a teacher tends to provide very few examples, or when a teacher "skip steps" in an example or proof so that you are unable to follow how some steps connect to the next?

When you know that you don't understand something about the notes during class, you should be ready and willing to ask your teacher questions. You should not wait to deal with your confusion until after you get home unless your difficulty is so minor that it won't prevent you from following the rest of the class notes. Do not be inhibited from asking questions in class. If asking questions does nothing else for you, the sound of your own voice as an alternative to the teacher's should at least keep you awake!

A question that you might ask could be as general as:

"How does the fourth line on the board follow from the third?"

If the teacher's explanation leaves you puzzled or unsatisfied, ask for further clarification. Persist until you are given an answer that is clear to you. When you finally have a satisfactory answer to your question, make sure to write the clarification into your notes even if your teacher does not add it to the board notes.

What should you do if a topic seems so difficult or the teacher's pace so fast that you know that you don't understand, but you can't even formulate an appropriate question? You might be tempted to throw up your hands and say to your teacher, "I'm lost!"

When faced with this situation, look at the board notes and identify the last line that you *did* understand. Then ask your teacher how the next line follows from that one. For example, suppose there are ten lines or steps written on the board, perhaps a collection of equations and statements. It occurs to one student, Aimee, that she is totally confused and doesn't feel that she understands any of the board work.

Aimee then looks at the first few steps and realizes that she did follow steps 1, 2, and 3 but that from then on she was lost. What Aimee can do is to ask her teacher:

> "I am confused back at step 4. Would you explain how step 4 follows from step 3?"

Once Aimee obtains sufficient clarification of how step 4 follows from step 3, she may then realize that she easily sees how step 5 follows from step 4 and step 6 from step 5, but she does not see how step 7 follows from step 6. She then can ask her teacher:

> "Would you explain how step 7 follows from step 6?"
>
> or
>
> "How does that step follow from the one above it?"

**Whenever you feel "lost," ask your teacher to explain the first step that you did not understand; then question any later steps that you still do not follow.**

If you tend to ask a lot of questions in class, you may want to limit the number you ask in order to give other students a chance. Again, if you hold back any questions, write them down so you can deal with them later.

Remember to add all helpful teacher (or student) comments and clarifications to your notes during class; if you wait to do this until you get home and review your notes, you run the risk of forgetting what you wanted to add.

Sometimes in class you may have a different problem understanding your notes. You think that you follow all the steps; you are not confused and you have no questions. You just do not think that you have a good idea of the overall picture of what the day's class notes are about. You may not see the goal of the notes or why the steps proceeded in the direction that they did. If you have this kind of difficulty, you might ask your teacher questions like:

> "What is the goal of this procedure?"
>
> "Why did you use that strategy in the procedure (or proof, example, and so on) instead of some other?"

| **When you cannot see the overall picture or goal of what the teacher is doing, ask questions.** |
|---|

## USE AN AUDIOCASSETTE RECORDER

Some students benefit if they listen to more than one presentation of the notes by their own teacher.

If your teacher allows it:

> **Consider using an audiocassette recorder to help you master your class notes, but never use it as a substitute for taking class notes.**

Even if you do record a class, taking good notes is still important, since unless you use a *video*cassette recorder you won't have the chance to see any pictures, diagrams, tables, graphs, or written notes that were not verbally stated by the teacher. So take the best notes you can, even if you do use a recorder. Your tape recording can then be used later, at the beginning of your study period. When you begin your studying, read through your class notes. Then listen to your recording with your notes in front of you. This will allow you to write in clarifications and corrections to your notes at your own pace, since you can stop the recording and play it back whenever you like. Remember, though, that an audiocassette recording is no substitute for taking good class notes and paying attention in class.

## WORK EFFECTIVELY IN A COOPERATIVE LEARNING CLASS

As part of the trend in education toward more active learning, cooperative or collaborative learning classes are becoming more common in math courses. In these classes, students are organized into groups of several students each. The teacher, who may use the lecture method only occasionally, mainly acts like a "coach" to the different groups of students as they work on problems that the teacher has assigned. In some of these courses, each class group is also responsible for meeting outside of class to work on problems together.

The cooperative learning style can have certain advantages over the traditional approach for some students. For example, the cooperative learning approach can:

1. provide opportunities for students to actively ask questions when they don't understand.
2. allow students to receive explanations from peers rather than from a "teacher."
3. encourage students to explain concepts to other students, an activity that can increase their self-confidence.

In some situations, you may have the option of choosing to register for a math course that is organized along the cooperative learning style described above, or for the same course taught by more traditional methods. How should you make your selection? Ask yourself these questions:

1. Am I comfortable working in groups?
2. Do I strongly prefer to work alone in and out of class?
3. Would I prefer to spend class time sitting and listening to the teacher while taking notes rather than working on problems together with other students?

**If you are unsure which type of course you prefer, it could be instructive for you to take a cooperative learning class.**

All of the suggestions for success in this book are equally important in cooperative learning, but there are some special problems you may encounter in such courses. Two of the most likely are the following:

1. You are dedicated to your group work, but other students in the group do not focus on the group's work during class time. Their minds may wander, they may just want to socialize, or for whatever reason they may not want to do the work.
2. You may feel that you have learned some concepts more slowly (or more quickly) than others in your group. You may worry that your slower learning will hold others back or that if others are slower than you, they will hold you back.

The fastest way to deal with the first problem is to raise your hand and ask a question that causes the teacher to come over to your group. Your teacher's presence (and the teacher's response to your question) will likely get the entire group back to working on the assigned problems.

To deal with the second issue, you need only change your attitude. If you have learned a concept more quickly than others in your group, you have to develop the interest and patience to explain such concepts to group members who do not yet understand. As mentioned elsewhere in this book, the act of providing such explanations will deepen your own understanding of the concepts and increase your self-confidence. If you are learning some concepts more slowly than others, you should not feel uncomfortable about asking for help from those who already understand. You will benefit from receiving explanations from your peers, and they will benefit from giving the explanations. As they help you, you are helping them.

## SUMMARY

### How to Use Class Time Effectively

Try to answer the question on the left before you look at the answer on the right.

|  |  |
| --- | --- |
| 1. What should be my main goal during class? | Learn as much as possible during class time. Take as complete a set of notes as possible. |
| 2. When should I ask questions? | I should feel free to ask questions in class whenever I don't understand something. |
| 3. How often should I attend class? | Attend all classes, and arrive on time or a little early. |
| 4. What should I do if I miss a class? | Obtain a clear, accurate copy of the notes, the homework assignment, and any handouts that I missed. Make a significant effort to work out the homework problems before the next class. |
| 5. How can I receive an extra benefit from an instructor who teaches more than one section of my course? | I can attend two sections of the teacher's course if possible. |

| | |
|---|---|
| 6. Where should I sit in class? | Sit toward the front and center of the room. |
| 7. How should I organize my notebook? | Use a large notebook split into two equal parts—one for class notes, one for homework. |
| 8. How should each day's notes be labeled? | Write the date at the beginning of each day's notes. Make sure to include any topic or chapter headings the teacher provides to help identify notes. |
| 9. How and where should I add to my notes? | I should leave a two-inch margin on the left side of a page, or I should write my notes on just one side of a page. I can add comments, examples, or clarifications during my studying. If I use a loose-leaf notebook, I can add supplemental pages to my notes at any time. |
| 10. What should I do in class if I can't follow how one step on the board connects to another? | Ask a question. If I cannot think of a more specific question, I might ask: "How does step $x$ follow from the previous step?" |
| 11. What should I do if I feel "lost" in class? | Identify the last step that I did understand, and ask questions that will clarify the steps from that point on. |
| 12. What should I do if I follow the details but do not understand the goal or direction of the notes? | Ask the question: "What is the goal of the procedure (proof, example, and so on)?" or "Why is this procedure (proof, example, and so on) done that way?" |

| | |
|---|---|
| 13. What should I do when I get answers to my questions in class? | Write the answers into my notes during the class. |
| 14. How can I benefit from using an audiocassette recorder? | Use an audiocassette recorder to record what happens in class. This tool should be used only as a supplement to my class notes. |

## PERSONAL WORKSHEET

I. Check the top five changes that you plan to make in using class time effectively in math courses. Add any other strategies you plan to use.

___ 1. I will ask questions when they occur to me in class.

___ 2. I will persist in getting answers to my questions by going to the teacher, the textbook, review books, or other students.

___ 3. I will miss a class only if I have an exceptionally good reason.

___ 4. If I have to miss a class, I will get the homework assignment and any handouts that the teacher may have provided.

___ 5. I will always arrive at class on time or even early.

___ 6. I will start thinking about class issues before the teacher begins to talk.

___ 7. I will get a seat in the front and center of the classroom.

___ 8. I will take a complete set of notes and add any helpful clarifications to my notes that I hear in class.

___ 9. I will take notes on only one side of my notebook pages.

___10. I will leave space in the margin to add clarifications to my notes.

**Other:** *I will* ______________________________

______________________________

______________________________

______________________________

______________________________

II. Rank the strategies you checked in section II according to how important you think they are for your success. Put the number 1 to the left of the most important, the number 2 to the left of the second most important, and so on.

III. After you try these strategies, refer back to this page and highlight the ones that worked best for you.

# CHAPTER 5

# How to Make Effective Use of Your Time Between Classes— Notes and Textbook

*Note: If you did very well on questions 6–11 of the "Classroom and Homework Habits" section of the diagnostic survey, you might choose to skip this chapter.*

This chapter and the next offer suggestions for how you can effectively use the time between classes to help you succeed in math courses. In order to make the best use of all the suggestions, you first need to organize your time efficiently.

## PLAN YOUR COURSEWORK STUDY SCHEDULE CAREFULLY

Many students plan their school activities on a day-to-day basis, but you will keep yourself better organized if you plan your study schedule ahead of time. A "month-at-a-glance" calendar, planner, or appointment diary book can be very useful for seeing the big picture and judging how your time should be planned during the semester or quarter. With this type of calendar, you can see an entire month of days on one page or two facing pages. There should be enough room in the space for each day for you to write down any appointments or important activities for that day. Using this type of calendar can provide you with a clear overview of your plans for the month or for several months.

For short-term planning of your activities, insert a "week-at-a-glance" calendar in the same book as your "month-at-a-glance" calendar. The weekly calendar allows you to detail your activities for each day of the week and to see all your plans for the week on one spread. You can write down your plans for the next day and never be unsure of what you need to do.

What's the best way to use these calendars? In general, the planner or appointment diary will contain your "to do" lists. The day before, write down a list of all your responsibilities for the next day, including study times for math and other subjects. Include personal responsibilities that are unrelated to coursework. Your list should also contain the estimated number of hours allotted for each activity.

List the things you have to do each day in order of importance. Although math does not have to be the first item on your list, it should be high enough so that you are sure to get to it. For example, you might write down the following schedule for Sunday and Monday:

| *Sunday* | *Hours* | *Monday* | *Hours* |
|---|---|---|---|
| Math | 3 | History | 2 |
| English | 2 | Math | 2 |
| Laundry | 2 | Science | 1 |
| Read Sunday paper | 1 | Work at supermarket | |
| Clean room | 1 | | 4 |
| | 9 | | 9 |

When you sit down to begin your work, you only have to look at your list to find something useful to do.

**Accomplish more in less time by listing all of your responsibilities for each day before that day starts.**

It is usually smart to allot more time than you think you will need for each activity. For example, if you think you can probably finish your math homework in a little over an hour, allot two hours on your list for this activity. Then, if you actually finish your math homework in less than two hours, not only can you gain a boost to your morale by crossing that task off your list, but you also get to cross off more hours than you actually used.

It is important, however, not to allot more hours than you actually have available. For example, if you have allotted nine hours for Monday, as listed in the sample schedule, make sure you have at least that many hours available on Monday to perform the tasks on your list.

Remember, too, that the time of day you pick to study math can be important.

**Choose a time of day to study math when you are especially alert.**

## PREPARE FOR YOUR NEXT MATH CLASS

Preparing for your next math class may involve several related activities. You might:

1. Read your class notes.
2. Read the relevant textbook sections.
3. Work out homework problems.

When many students do what they call "homework," they go through the motions of quickly reading through their notes, maybe glancing at their textbook, and working out homework problems as quickly as possible with as little thought as possible. For these students, this kind of effort is sufficient to make them feel that they are working hard in the course. Unfortunately, they are only kidding themselves.

You can do a much better job than this to prepare for your next class. This chapter tells you specifically how to deal with your class notes and your textbook. Chapter 6 tells you how to handle your homework assignments and related issues.

## GET HELP OUTSIDE CLASS

If you still have problems understanding your notes at the end of the class period, you should feel free to visit your teacher during his or her office hours. Many teachers are receptive to helping a student in this environment where they can more easily focus on one person at a time. Furthermore, even teachers who are difficult to follow in class may be very helpful on a one-to-one basis. Don't give up on your teacher without giving those office visits a chance.

Some students are uncomfortable visiting teachers in their offices. If you feel this way, it may help if you find a classmate or two to go along with you. You may benefit from the teacher's response to your classmates' questions even if you had not thought of the same questions yourself.

What should you do if your teacher is unavailable or unwilling to help you, or if, for any reason, you don't want to ask your teacher for outside help? If this is the case, try to find another teacher who is willing and able to answer your questions about the course. Whoever helps you, be sure to ask your questions in a friendly and courteous manner and show appreciation for the teacher's time.

In addition, other students in the class or those who have already taken the course might be willing to help you if you ask. Try to develop relationships with classmates and other students so that you will feel comfortable asking them questions about the course.

To derive the most benefit from teachers or students outside class, put question marks in your notes to remind yourself what you need to ask about. You should also write down a list of specific questions that you plan to ask before meeting with the person who is going to help you. The act of creating questions to ask will make it easier for you to benefit from the session. Be sure to take notes during your help session. Finally, use your notes to review what you learned as soon as possible. If you wait too long to review or don't review at all, you will likely forget much of what you learned in the help session.

Many colleges have a learning assistance center or math lab to help students. These will be discussed in Chapter 13.

> **To get the most benefit from teachers or other students outside class:**
>
> **1. Use question marks to identify confusing material in your notes or textbook.**
>
> **2. Write down specific questions that you will ask.**
>
> **3. Afterward, be sure to review what you learned at the help session so you don't forget what you learned there.**

## READ YOUR MATH CLASS NOTES AND TEXTBOOK EFFECTIVELY

If you have a free period after your math class, spend some of that time reviewing the notes you took during that class. At this time, your notes will be as fresh in your mind as they are ever going to be. This is a good time to add any explanation or missing information to your notes. If you don't have a free period, at least read over the notes later on that same day. Even five or ten minutes of review time can be a big help.

> **Read your math notes on the same day that you wrote them.**

Some very dedicated students rewrite their class notes in another notebook once they are at home. Because this allows you to write at your own pace, these notes can be neater and better organized than the notes you took the first time. Furthermore, this act of rewriting can deepen your understanding of the concepts presented, as well as enhance your ability to remember them. Some students find that printing instead of "writing" the notes the second time slows them down enough that their comprehension improves.

You might find it beneficial to highlight or underline the most significant formulas, equations, definitions, theorems, and so on, in your notes or textbook. The process of choosing what to highlight or underline contributes to your being a more active learner. In addition, highlighting or underlining helps you to focus on the key points when you are reviewing. However, do not overdo it. Highlighting or underlining too many things will destroy the value of the process.

## USE ADDITIONAL TECHNIQUES TO IMPROVE YOUR NOTES

How you read and develop your notes depends on what type of notes you have. Notes in a math course typically consist of collections of some of the following:

- a statement or discussion of a definition
- a statement of a theorem
- an example or a discussion of examples
- a description of a procedure for solving a certain type of problem
- a proof of a theorem or a derivation of a formula
- a list of steps of a procedure
- a list of formulas or equations

Read your notes and textbook with pen or pencil in hand. You should first identify and label each section of your notes and textbook according to which of the categories it belongs to. The following paragraphs describe how you should treat a section of your notes if it includes any of the items on the list.

## A Statement or Discussion of a Definition

If you are reading a definition or a discussion of a definition:

1. Think of examples that illustrate the definition and write them in the margin or on the facing page to the left of the definition or discussion of the definition.
2. Think of examples that violate the definition, and describe why they do. Write these examples in your notes in the margin or on the facing page to the left of the definition or discussion of the definition.

**Example**

| *Additions to Notes* | *Notes* |
|---|---|
| 1. $3x^2 + 4x + 5 = 0$ is an example of a quadratic equation.<br>2. $5x - 3 = 0$ is an example of an equation that is not quadratic, since the highest power is not $x^2$.<br>$5x^3 - 2x^2 + 3x - 4 = 0$ is not quadratic, since there is a third degree term. | Definition: A *quadratic equation* is an equation that can be written in the form $ax^2 + bx + c = 0$ where $a$, $b$, and $c$ are constants with $a \neq 0$. |

If your notes or textbook already contain such examples, try to think of new ones to include in your notes.

## A Statement of a Theorem

If you are reading a statement of a theorem:

1. Think of examples that illustrate the theorem. Write these examples in your notes in the margin or on the facing page to the left of the statement of the theorem.
2. Think of examples that do not satisfy the conclusions of the theorem because they do not satisfy at least one of the assumptions. Write these examples in your notes in the margin or on the facing page to the left of the statement of the theorem.

### Example

| *Additions to Notes* | *Notes* |
|---|---|
| 1. Examples of theorem<br>$x^2 \cdot x^3 = x^5$<br>$x^7 \cdot x^4 = x^{11}$<br>$3^5 \cdot 3^8 = 3^{13}$<br>2. Examples of when theorem does not apply<br>$x^2 + x^3 \neq x^5$<br>The theorem applies only to the *product*, not the sum of two terms with the same base. | Theorem: If $n$ and $m$ are positive integers, then $x^n \cdot x^m = x^{n+m}$ |

If your teacher has already included such examples in the notes, try to think of other ones to include.

## An Example or a Discussion of Examples

If you are reading an example or a discussion of examples, state the point of each example:

1. Identify what topic, concept, definition, theorem, or formula it illustrates. Write down such points in the left margin of your notes.
2. Identify how you would recognize the point the example illustrates from the way the example is written. Write down such points in the left margin or on the facing page.

**Example**

| *Additions to Notes* | *Notes* |
|---|---|
| 1. Example of solving a linear equation involving fractional terms.<br>2. The method involves multiplying the equation by the least common denominator of 2, 3, and 4. | Example: Solve for $x$:<br>$\frac{x-2}{4} - \frac{x}{3} = \frac{1}{2}$<br>Solution: First multiply all three terms by 12.<br>$12\left[\frac{x-2}{4}\right] - 12\left[\frac{x}{3}\right] = 12\left[\frac{1}{2}\right]$<br>$3(x-2) - 4x = 6$<br>$3x - 6 - 4x = 6$<br>$-x = 12$<br>$x = -12$ |

## A Description of a Procedure for Solving a Certain Type of Problem

If you are reading such a description, answer questions like these:

1. What kinds of problems can you apply this method to? Write down your answer.
2. What are any key points that you should remember about the method? Write down your answer.
3. What are any important differences between this method and other methods that apply to similar, but different problems? Write down your answer.

**Example**

| *Additions to Notes* | *Notes* |
|---|---|
| 1. Problem must have an $x^2$ term.<br>2. Equation must be set equal to 0. e.g.; $4x^2 = 3x - 5$ must be changed to $4x^2 - 3x + 5 = 0$ before the quadratic formula can be applied.<br>3. Method applies to quadratic equations. Often, there are two answers for $x$. If:<br>$b^2 - 4ac > 0$, there are two real answers.<br>$b^2 - 4ac = 0$, there is one real answer.<br>$b^2 - 4ac < 0$, there are no real answers. | To solve a quadratic equation, $ax^2 + bx + c = 0$, with $a \neq 0$, use the quadratic formula<br>$x = \dfrac{-b \pm \sqrt{b^2 - 4ac}}{2a}$ |

## A Proof of a Theorem or a Derivation of a Formula

If you are reading a proof of a theorem or a derivation of a formula, answer the following questions:

1. Can you see how each step in the derivation or proof of the theorem follows from the previous step? Add any comments.
2. What theorems or definitions are used in the derivation or proof? Write them down.
3. What mathematical knowledge used in the derivation or proof might be useful to remember? Write it down.

4. What is the formula used for? Write it down.
5. In what ways, if any, would you be likely to use the formula at the wrong time?
6. Write into your notes any insights you have gained from answering these questions.

**Example**

*Additions to Notes*

1. The 7th line follows from the 6th by taking the square root of each side of the equation. The $\pm$ sign is on the right, since there is both a positive and a negative square root.
2. The last step follows from the definition that allows us to combine two fractions with the same denominator.
3. Dividing the equation through by $a$ in line 2 makes it easier to complete the square in line 4. That $a$ is not 0 allows us to divide through by $a$.
4. Formula is used for solving the quadratic equation $ax^2 + bx + c = 0$ where the left side is not easily factored.

*Notes*

If $ax^2 + bx + c = 0$, where $a$, $b$, and $c$ are real numbers, $a \neq 0$,

$$x = \frac{-b \pm \sqrt{b^2 - 4ac}}{2a}$$

Derivation

$$ax^2 + bx + c = 0$$

$$x^2 + \frac{bx}{a} + \frac{c}{a} = 0$$

$$x^2 + \frac{bx}{a} = -\frac{c}{a}$$

$$x^2 + \frac{bx}{a} + \left[\frac{b}{2a}\right]^2 = -\frac{c}{a} + \frac{b^2}{4a^2}$$

$$\left[x + \frac{b}{2a}\right]^2 = \frac{-4ac}{4a^2} + \frac{b^2}{4a^2}$$

$$\left[x + \frac{b}{2a}\right]^2 = \frac{b^2 - 4ac}{4a^2}$$

| | |
|---|---|
| 5. If the quadratic can be factored, such as in the example, $x^2 - 2x - 3 = 0$, we do not have to use the quadratic formula to solve for $x$. | $x + \frac{b}{2a} = \frac{\pm\sqrt{b^2 - 4ac}}{2a}$<br><br>$x = -\frac{b}{2a} \pm \frac{\sqrt{b^2 - 4ac}}{2a}$<br><br>$x = \frac{-b \pm \sqrt{b^2 - 4ac}}{2a}$ |

## A List of Steps of a Procedure

If you are reading such a summary list, answer the questions:

1. What does this procedure accomplish? Write it down.
2. What is the reasoning behind the order of the steps? Write it down.

**Example**

| *Additions to Notes* | *Notes* |
|---|---|
| 1. How do you find the equation of a line in slope-intercept form when two points on the line are given?<br><br>2. Finding the slope in step 1 must be accomplished first before step 2. | Finding the equation of a line in slope-intercept form passing through the points $(x_1,y_1)$ and $(x_2,y_2)$ in slope-intercept form.<br><br>Step 1: Find the slope $m$.<br><br>$m = \frac{y_1 - y_2}{x_1 - x_2}$<br><br>Step 2: Substitute $m$ into the equation.<br><br>$y - y_1 = m(x - x_1)$<br><br>Step 3: Solve for $y$ to put the equation in the form<br><br>$y = mx + b$ |

## A List of Formulas or Equations

If you are reading such a list, for each formula or equation, answer the following questions.

1. What does it do?
2. How do you decide when the formula should be used?

**Example**

| *Additions to Notes* | *Notes* |
|---|---|
| First formula:<br>1. Tells you when to add exponents when combining algebraic terms.<br>2. It is used when multiplying terms with the same base, e.g., $3^2 \cdot 3^4 = 3^{2+4} = 3^6$ | Rules of Exponents<br>I. $a^n \cdot a^m = a^{n+m}$<br>II. $(a^n)^m = a^{nm}$ |
| Second formula:<br>1. Tells when to multiply exponents when combining algebraic terms.<br>2. It is used when taking a single term to a power in parentheses with another exponent on the outside, e.g., $(3^2)^4 = 3^{2 \cdot 4} = 3^8$ | |

If you treat your notes and textbook in the ways just described, you will derive the greatest benefit from them. The point is to become an *active participant* in the creation of your notes. Following this process will naturally increase both your understanding of the concepts and your memory of them.

## USE ADDITIONAL TECHNIQUES TO READ YOUR TEXTBOOK

Reading a math textbook is not like reading a romance novel. A math textbook needs to be read slowly with pen or pencil in hand. The textbook does not, however, have to be read page by page, in order. For example, many chapters and sections have introductory material that you might better understand after you read the examples that follow the introductions.

Do not even start to read a textbook section unless you have at least a general understanding of the relevant material in the sections that precede it. In addition, you should have a sufficient knowledge of any relevant concepts in preceding chapters.

How should you read a textbook section? Whenever you are reading a discussion, an explanation, a definition, a theorem, a proof, or any other abstract material that confuses you, stop reading. Then reread the confusing material to try to clear up your confusion.

**Place a question mark next to the confusing material. Reread the material until it is clear to you.**

Often you will need to read the examples that follow an explanation to clarify any lack of understanding. After reading through a textbook example, try to rework the example before looking back at the solution. If you have difficulty in doing this, examine the solution for help. Then try again to rework the example, again without looking at the solution.

**To deepen your understanding of an example, rework the example without looking at the solution.**

In addition to the above advice, you can use the same system for writing in your textbook that was described for adding notes to your class notes.

When you begin to use your math textbook, you should read the introduction to the book to determine the features included at the back of each chapter. These features are intended to provide you with additional help in mastering the material in the chapter. For example, if there is a glossary, you should use it to help you check the definitions of key terms as you are reading the chapter. You should also use the glossary after you have finished the chapter to review the definitions.

Another feature provided at the end of each chapter of most math books is "review problems" or "chapter exercises." If your book contains such problems, you should attempt to work out many of them, but only after you have worked out many of the problems in each section. Unlike problems at the end of each section, the chapter exercises test your ability to work out problems without knowing ahead of time what concepts are being tested. Such practice is extremely useful, especially when studying for tests. This idea is discussed further in Chapter 10.

In addition to chapter exercises or review problems, some math books have "self-tests," "review questions" or "study questions" at the end of each chapter. If your book has any of these features, be sure to try to answer these questions after you have read the chapter. Your efforts to complete these end-of-chapter questions will definitely increase your understanding of the concepts.

**Make use of helpful features at the end of each textbook chapter to help you solidify your understanding of concepts and your ability to work out problems in the chapter.**

## READ OTHER TEXTBOOKS

What should you do if your textbook is often extremely difficult to follow?

> **Use additional textbooks as resources to help you in the course.**

Ask your teacher to recommend additional books to you. Teachers are usually happy to make such recommendations, since your request demonstrates interest in the course. Furthermore, teachers are likely to be willing to lend you one or more textbooks from their usually overstuffed bookshelves.

Other sources of books and study guides are:

- your school library
- your school's learning assistance center
- other school libraries
- public libraries
- your school bookstore
- other bookstores

For a list of textbook recommendations for a variety of math courses, see Appendix B at the back of this book.

Even if you are happy with your textbook, it can be useful to have these additional resource materials on hand when you do your homework and prepare for tests.

For example, one student, Nicole, tried to read the relevant sections in her textbook to help her understand the class notes. Because she had great difficulty understanding her textbook, she went to her school library to search for some other textbooks that dealt with the same material. She found the math section where these other textbooks were located. By examining the index or table of contents of one book at a time, she was able to locate the topics she was interested in. In a short time, she found two books that had clearer explanations of the relevant topics than she had found in her own textbook. She checked these two books out of the library so that she could refer to them when she wanted additional explanations to help her understand upcoming topics in the course.

## SUMMARY

### How to Make Effective Use of Your Time Between Classes—Notes and Textbook

Try to answer the question on the left before you look at the answer on the right.

|  |  |
|---|---|
| 1. How can I do more work in less time? | I should list all my responsibilities for each day before I start. |
| 2. What should I do if I need clarification of my notes or textbook outside of class? | Use question marks to indicate confusing material in my notes or textbook. Write a list of specific questions to ask my teacher, another teacher, or other students. |
| 3. When should I read my notes? | On the day that I write them. |
| 4. How and where should I expand my notes? | While reading, expand my notes by using the left margin or the facing page to add examples or questions, depending on the nature of the notes; for example, a definition, an example, a derivation, a theorem, a procedure, and so on. |

| | |
|---|---|
| 5. How should I read my textbook? | Highlight or underline important concepts. Place a question mark next to confusing material and then find examples to study. When the examples are clear, reread the material to try to answer my question. |
| 6. What are some end-of-chapter features that will help me master concepts in the chapter? | The backs of chapters may contain glossaries, self-tests, or chapter exercises. |
| 7. What other reading material can I find to help me in the course? | I can find additional textbooks to help me master concepts. |

## PERSONAL WORKSHEET

I. Check the appropriate column for each statement.

| | always | usually | often | sometimes | never |
|---|---|---|---|---|---|
| 1. I list my responsibilities for each day before I begin working on them. | | | | | |
| 2. I read my notes before the next class. | | | | | |
| 3. I visit the teacher during office hours. | | | | | |
| 4. When I visit the teacher during office hours, I bring a list of specific questions to ask. | | | | | |
| 5. I make time to review what I learned during the teacher's office hours. | | | | | |
| 6. I mark confusing material for later reference. | | | | | |
| 7. I reread notes and text material and examples that follow. | | | | | |
| 8. I rework examples without looking at the solutions. | | | | | |
| 9. I use additional textbooks to help me in the course. | | | | | |

II. Look at the statements in Part I that do not have a check in the "always" column. Circle the numbers of the strategies that you plan to follow more often.

## PRACTICE

1. If you are currently in school, buy a calendar or planner. Use the planner this weekend to
   a. Write down a list of your goals for the upcoming week without specifying your goal for each day. Include items like studying for a test, reading chapters of books for any course, writing papers, and personal items such as laundry or food shopping.
   b. Write down your goals for Monday, making sure that completion of these goals will bring you closer to meeting the goals for the week that you listed in step a.
2. If you are currently in a math course, go to a library and find several textbooks that cover the same subject as the one assigned for your course. In each textbook, look up three sections that you found difficult to understand in your own textbook. Check out and take home the one or two textbooks that you found to be the clearest.

# How to Make Effective Use of Your Time Between Classes—Homework and Beyond

*Note: If you did very well on questions 6–11 of the "Classroom and Homework Habits" section of the diagnostic survey, you might choose to skip this chapter.*

There are many things you can do between classes besides working with your notes and textbook.

## DO YOUR HOMEWORK BEFORE THE NEXT CLASS

Obviously, you must do the homework if you are to master the concepts. But does it really matter *when* you do it?

**The best time to do your homework is before the next class session and, ideally, on the same day it was assigned.**

This is the time that the explanations from the class will be freshest in your mind.

If you can't do the homework immediately, then at least make sure to do it before the next class session. Even if you have trouble solving some of the problems, do as many as you can. If you establish this habit, you can master the course with the least possible effort, since it is easier to study and absorb concepts in smaller portions than it is to allow homework from several classes to accumulate.

This advantage of being able to deal with smaller portions of homework and notes applies not only to math but to most other subjects as well. Though it is never desirable to miss a class, missing a class in a *cumulative* subject such as math can be disastrous. In this kind of course, in order to understand what is going on in class, it is often necessary to understand what happened in previous classes, and especially in the most recent one.

If you cannot do most or all of the homework, make a dedicated effort to get help from another student or your teacher *before the next class*.

If you do not do the homework before the next class, you have created two problems for yourself. First, it will now be more difficult for you to do that homework and understand the covered material; second, you may make it more difficult to understand the material that is covered in the next class. So, while doing homework in preparation for your next class, you may actually be preparing for future classes as well.

## WORK OUT HOMEWORK PROBLEMS WITH CARE AND PERSISTENCE

It is impossible to give you specific advice on how to solve problems without referring to particular problems. But here are some general principles for dealing with homework problems:

1. Write down a summary of the information you are given.

2. Write down a description of what you need to find.
3. Write down any formulas or theorems that may help.
4. If possible, begin by trying to give an estimate of the right answer. Then when you do work out the problem, compare your answer with your estimate to see how closely they match.
5. See if similar examples worked out in your notebook or textbook might help you.
6. If you are having difficulty working out a problem, try to apply any remaining given information that you have not used. However, remember that some problems may contain more information than is needed to solve the problem.
7. Persist in working on what seems like a difficult problem, but do not spend so much time that you neglect the rest of your homework and other responsibilities. One good approach is to leave a

problem you are having great difficulty with, go on to some other problem or activity, and later come back to the problem.

8. Check your answers by looking at the back of the book or by asking other students or your teacher. However, *do not look at any answers or get help before making a significant attempt to work out the problem yourself.*
9. After finishing a day's worth of math homework problems, summarize the main procedure or steps in your own words. Also, identify any typical errors that a student should avoid.

## CHECK YOUR HOMEWORK BEFORE THE NEXT CLASS

Ideally, you should check your homework as soon as possible after you do it. In any case, it is important to find a way to:

> **Check your homework solutions before the next class.**

Solutions may be found in the back of your book or on answer sheets given by your teacher. In addition, you can consult with other students.

There also are several things you can do yourself to check your homework. Any one or more of these approaches might work, depending on the type of problem:

1. Reread the problem to make sure you didn't miss anything.
2. Estimate the right answer before you work it out. Verify that your answer is close to your estimate.
3. Substitute your answer back into the problem to see if it fits the given information.

4. Redo the problem in the same way.
5. Redo the problem in a different way to see if your answers match.

It is neither necessary nor desirable for you to apply all of these procedures in checking any one problem. In addition, not all of these procedures would work with every type of problem. For example, you can always redo a problem in the same way, but you might not be able to find a different way. Certain problems may require different checking procedures. Therefore, be sure to ask your teacher for suggestions on how to check solutions for the kinds of problems you are studying.

It might help to see examples of checking problems using some of the approaches suggested.

## Estimating the Right Answer

Suppose you want to solve for $x$ in the equation

$$4.8x - 12 = 31.2$$

One approach to estimating the solution before solving is to round off 4.8 to 5, 12 to 10, and 31.2 to 30 and solve the simpler equation

$$5x - 10 = 30$$
$$5x = 40$$
$$x = 8$$

Therefore, $x = 8$ is a reasonable estimate of the answer to the original equation. Now, solving the original equation directly, you have

$$4.8x - 12 = 31.2$$
$$4.8x = 43.2$$
$$x = \frac{43.2}{4.8}$$
$$x = 9$$

Since your solution $x = 9$ is close to your estimate $x = 8$, you have some evidence that you are correct. If you had mistakenly obtained $x = 90$, you should have suspected that this answer was wrong since 90 is so far from your estimate of 8. Estimating is not a foolproof way of verifying that your answer is correct, but it can help you rule out many wrong answers.

## Substituting Your Answer Back into the Problem

Suppose, again, that you are solving for $x$ in the equation

$$4.8x - 12 = 31.2$$

and you obtain the answer $x = 5$. To see if $x = 5$ fits the given information, substitute the number 5 into the given equation for $x$:

$$4.8(5) - 12 = 12 \neq 31.2$$

Therefore, $x = 5$ is incorrect. Of course, the correct answer is $x = 9$. This solution can be checked:

$$4.8(9) - 12 = 31.2$$

## Redoing the Problem in a Different Way

This approach can be illustrated in working the following multiplication problem:

$$(x + y)(x + y)$$

Here is one method of working out this problem:

$$\begin{aligned}(x + y)(x + y) &= (x + y)x + (x + y)y \\ &= x^2 + yx + xy + y^2 \\ &= x^2 + 2xy + y^2\end{aligned}$$

Now, this happens to be the right answer. But let's see if the problem can be checked by solving it in a different way:

$$
\begin{array}{ccccc}
x & + & y & & \\
x & + & y & & \\
\hline
 & & xy & + & y^2 \\
x^2 & + & xy & & \\
\hline
x^2 & + & 2xy & + & y^2
\end{array}
$$

Since you found the same answer using a different method of multiplying, you can be more confident that your answer is correct.

For some problems, redoing the problem in a different way or even estimating the right answer may be difficult or impossible. For such problems, the only checking activity that you can use without the help of answer books or other people might be to redo the problem in the same way.

Even if you can use the answer key in your textbook or find other people to help check your homework, checking the homework yourself will produce several benefits:

1. You will deepen your understanding of the concepts involved in the problems through the process of finding and applying checking procedures.
2. You will gain excellent practice in depending only on yourself for checking problems. This will help you during tests when you have no alternative way of verifying your answers.
3. You will increase your confidence in your ability to succeed in the course.

As you can see, the process of checking your work is a very important step. If your answers are correct, great! But if you discover any wrong answers, try reworking the problem until you get the right answer. When you finally get the right answer, think about why you were wrong the first time. Analyzing your difficulty now can reduce your chance of making similar errors later on tests.

Of course, sometimes an answer in the back of the book is incorrect. It is also possible that both the book's answer and your answer are correct, but they look different. If your answer does not agree with the one in the back of the book, one of the following possibilities is true:

1. You've made a mistake.
2. Your book is wrong.
3. You and the book are both right. In this case, one possibility is that both your answer and the book's appear to be different but are approximately equal. For example, if the correct answer to a question is 6.257, the book may approximate it as 6.26. Another possibility is that both answers are equivalent. For example, you may find that an answer to a problem is $2a + 2b$, but the book's answer may be the equivalent value of $2(a + b)$.

So, if you think your answer is correct, but it does not exactly match the one in the book, try to determine if the two answers are approximately equal or equivalent. If you can't do this, check with another student or your teacher to see which answer is correct: yours, the book's, or both yours and the book's.

## CONSTRUCT A LIST OF TEST TOPICS BASED ON YOUR HOMEWORK

**You can get a head start in preparing for the test for the unit by constructing a list of potential topics for the test as the unit is going on and not waiting for the period just before the test.**

(Chapter 8 will go into greater detail on how to construct a list of test topics.)

To build your list in your notebook, first reserve several pages at the beginning of your notes for the unit or at the back of your notebook. Begin to construct your list of topics after doing the first homework assignment in the unit. Then write down a list of the covered topics using the exercise headings in your textbook. For example, let's suppose that beginning on page 16 of your textbook there are thirty-five exercises containing quadratic expressions under the direction "Factor each of the following." Then on page 20, there are twenty-four exercises involving expressions with numerical exponents under the direction "Simplify the following expressions." Your list might look like this:

1. Factor quadratic expressions. (16/1–35)
2. Simplify an expression involving numbers with numerical exponents, e.g., $(4^2)^3$. (20/1–24)

In each case, the pages and problem numbers of the exercises that deal with these topics are placed to the right of that topic.

If these are the first two homework topics, you should begin constructing the list after completing the homework assignments. As the teacher continues to assign homework during the unit, you should keep adding topics to your list as soon as you have completed the homework. By the time of the test, you should have an organized list of topics for the unit. As the test time gets closer, you would do well to verify with your teacher that your list is complete.

**The best time to begin to construct your list of topics for a unit is after completing the first homework assignment. The list should be continued and completed on a step-by-step basis as you do homework in the unit.**

## REINFORCE YOUR KNOWLEDGE OF PREVIOUSLY COVERED CONCEPTS

On class days when you have little or no math homework, there is still something you can do that can help you greatly in the course.

> **Whenever you have the opportunity, review concepts from previous classes by reworking homework problems or working out new problems, one topic at a time, from your developing list of topics. Your goal is to become an expert on each topic before going on to the next.**

If you frequently reinforce your understanding of previous concepts in this way, it will make it much easier for you to prepare for your next test.

## USE COURSE OUTLINE BOOKS AND STUDY GUIDES

The education or course outline sections of many bookstores contain math "review" books or study guides designed for many specific math courses. These books outline a particular course. More significantly, they provide numerous worked-out examples that allow you to supplement the examples you have already seen in class, your textbook, or your homework. In addition, these books often provide summary collections of problems and answers that you can use to help prepare for tests.

> **Work with a study guide or course outline book that applies to your particular course.**

Before you buy a particular book, make sure to read several examples or explanations in different books. Identify which book is the clearest to you and the most comprehensive for your purposes. In this way, you can best identify which book will be the most beneficial for you.

Here is a list of useful review book and college outline series that include math books:

- Schaum's Outline Series
- Harcourt Brace Jovanovich Outline Series
- AMSCO School Publications
- Barron's Educational Series

For a more detailed list of study guide recommendations, see Appendix B.

## WORK WITH OTHER STUDENTS

When you check your homework or miss a class, you should get help from other students in the class, either to check homework answers or to obtain class notes. Even when you are able to do your homework and have not missed a class, you still can benefit greatly from discussing class notes, asking questions of other students, and answering questions from them.

To determine which students would be best for you to work with, try to get to know other students in the class during the first two weeks of the course. In this way, you can identify those students with whom you might be the most compatible.

In addition to personal compatibility, you should find students who take the course seriously, who are likely to keep up with the homework and attend classes, and who seem to have better than average ability in the course. You should also try to identify students who might be willing to offer you help when you need it, as well as to receive help from you when they need it.

When you feel confident that you understand the work you are doing, encourage other students to ask you questions.

The act of explaining a concept to another student will deepen your understanding of the concepts. After you have correctly explained a concept to someone else, you have given yourself strong evidence that you understand that concept.

If you find several students with whom you are comfortable working, you might consider forming a study group that would meet on a regular basis. Discussing problems with a group of students can be more beneficial than discussing them with just one other student. Different students will have many different strengths in the course that can contribute to the knowledge of each student in the group. The cumulative strengths of the group can overcome many of the weaknesses of individual students.

**Work with other students in the course to pool your knowledge. Ask them questions about classwork and homework, and encourage them to ask you questions.**

In some math classes, particularly in cooperative learning classes, you may be *assigned* to work with a group of other students *outside of class*. One of the problems you may have in such a course is that one or more of the group or team members may not act responsibly. He or she may not keep up with assignments, may not contribute to group projects, or may miss group meetings. What should you do then if group members do not pull their weight? If you are the group leader, you should try to persuade the nonperforming students to start being more responsible. If that doesn't work, you need to report the facts to your teacher. If you are not the group leader, you should complain to your group leader about members who are not contributing. If the group leader does not solve the problem, then the problem should be reported to the teacher.

Another issue may occur when students are assigned to work in groups outside of class: Sometimes a student's final grade is affected by "group grades" that the teacher gives for group projects. In this situation, you should protest to your teacher if the grading system allows other students' lack of effort or achievement to possibly lower your own final grade.

There are ways for the teacher of such a class to avoid this problem. One effective way is for students' grades to be based on their own individual average in combination with "bonus points" they obtain from their group work, which may then improve their grade. If a student's group work average is lower than the average based on his or her individual work, then only the individual work will count toward the final grade. In this way, your participation in a group cannot hurt your own final grade.

**Assuming you have a choice, you should select a course involving group grades only when the teacher's grading system cannot penalize you for other students' inadequate work.**

## PREPARE QUESTIONS TO ASK IN THE NEXT CLASS

If you have questions based on class notes, the textbook, or your homework, write them down so that you may ask the teacher during the next class. If no questions about the homework or previous class notes occur to you, try to reread your notes and textbook, looking for potential questions.

The goal of looking for questions to ask is not to impress your teacher, but to clarify the concepts for yourself. The act of finding questions, or creating them, will enhance your comprehension of the concepts under discussion, as well as improve your memory of them.

How should you organize your collection of questions? List all your questions at the end of the notes from the last class. Try to make them as specific as possible. Here is a sample list of questions you might ask your teacher:

"I was reviewing the notes from the last class.
Would you give another example of a ___?"

"From yesterday's homework, what is the first step in doing problem ___ on page ___ of the textbook?"

"On page ___ of the textbook, why does line 5 follow from line 4?"

"What is the first step in solving this problem?"

"What is the next step in solving this problem?"

"Why do you solve this problem that way?"

"Would you give an example of that definition (theorem)?"

"What is an example of where the definition (theorem) does not apply?"

"What are some applications or implications of that definition (theorem)?"

"Why is this problem approached differently from the preceding one?"

**Make frequent efforts to ask questions. Do not allow a question you have from the homework problems or the previous class to go unanswered. Get help fast.**

Consider Lauren, who made a good effort on her calculus homework the previous night, but never could work out problem 5. In the next class, she doesn't ask her teacher any questions about the problem because she doesn't want to look "stupid" by letting the teacher and the class know that she could not do the problem.

In the same class, Michael also couldn't work out problem 5. But since he doesn't want to let the class go by without learning how to solve the problem, he raises his hand to ask the teacher for help. He is not worried that the teacher or the class will think that he is stupid. In fact, he believes other students probably had trouble with the problem and would appreciate his asking the question.

**Ask questions in class. Never avoid asking a question out of fear of looking stupid.**

## CREATE QUESTIONS TO ASK YOURSELF

There may be times when you don't *naturally* have questions about anything in the course. In this situation, you may begin to relax and not think about the concepts very much. However, rather than simply relax when no questions naturally occur to you, take a more active approach.

**Create questions about notes and homework to ask yourself and then answer.**

If you don't have any questions about concepts or a section, try to think of questions that you *might* ask. If you have trouble thinking of such questions, look at the list of sample questions in the last section. Try to write down your own questions in a style similar to those, but deal with the relevant concepts or section.

Nicole was a student in an algebra class who regularly did her homework in the course. Even though she could do the homework and no questions naturally occurred to her, she didn't feel confident about how well she understood the material. Therefore, she tried to create her own questions about the material.

First, she reread a definition in her textbook. Then she created the questions:

"What is an example of the definition?"

"What is an example where the definition does not apply?"

Then she looked at a rule (theorem) in her notes or textbook and created the questions:

"What is an example of where that rule (theorem) applies?"

"What is an example of where the rule (theorem) does not apply?"

"Why does this rule (theorem) follow from the previous one?"

"What use or uses does the rule (theorem) have?"

Next she looked at an example in her notes and created the questions:

"What does this example illustrate?"

"What are other examples of the same thing?"

Finally, she looked at a homework problem and created the questions:

"Is there another way to solve this problem?"

"Why don't we take this other approach to solving this problem?"

"How is this problem similar to or different from other related problems?"

After creating her list of questions, Nicole took some time to try to answer them. If after some effort she needed help, she referred to her textbook or classmates. If she still could not find a satisfactory answer to a question, she made a note to ask her teacher the question in the next class.

This was one of the most useful activities she could pursue in her attempt to master concepts in the course.

## MAKE PRODUCTIVE USE OF THE FEW MINUTES BEFORE CLASS

In order to get the most out of a class:

**Try working on course material for the few minutes just before the class starts.**

What can you do during this short period? If you have the time, try any of the following activities:

1. Locate in your homework the questions that you have written down to ask in class.
2. Reread the most recent class notes to find any last-minute questions and to prepare yourself mentally for the class that's about to begin.
3. Reread the section in the textbook covered in the last class or begin to read the one that the teacher will likely cover in the upcoming class.
4. Work out one of the unassigned problems from the section of the homework in the textbook.

These few minutes before class can be spent in the library or cafeteria, or in the classroom if it is empty and you can get there early.

## BREAK UP YOUR STUDY PERIODS

Research has shown that people are more likely to remember the first and last idea focused on in a study session than any other ideas studied during the session. It follows that the more separate study periods you have, the more "first" and "last" ideas you will focus on and remember. You may want to try the following strategy:

**Study in forty-five to sixty-minute segments with breaks that last between five and ten minutes.**

## COMPLETE COMPUTER HOMEWORK EFFECTIVELY

Some math courses involve computer assignments. Your teacher may assign one or two major projects or a number of smaller homework assignments requiring the use of a computer. In addition, you may have to work with computer software such as spreadsheets, word processing, or statistical packages. An assignment may require (1) working with the computer to produce appropriate computer output, and then (2) writing out and submitting an *interpretation* of that output to the teacher.

To produce the computer output, you should apply the following advice:

1. Do not be afraid of using the computer. If you have trouble knowing how to begin, get help from your teacher or from other students. Also, do not forget to use relevant software manuals or online help screens.
2. Begin working on the computer project as soon as it is assigned. Do not leave it for the last minute or you will put too much pressure on yourself.
3. Whatever help you get, make sure you *understand* any commands that you are inserting into the computer and why it produces the output that results. Do not yield to the temptation to mindlessly input commands into the computer without understanding what you are producing.

Ignoring this advice will likely increase your chance of making mistakes and make it more difficult for you to interpret your results intelligently.

After producing your computer output with care and understanding, spend some time (more than a few minutes) examining and analyzing your results. Then write out your interpretation, read what you wrote, and revise it. Try hard to use all of your numerical or graphical results in your interpretation. Keep revising your interpretation until it takes into account all your results and gives the clearest possible explanation of your findings.

You could quit now, but if you can make the time, take the opportunity to do more work on the computer assignment than your teacher required. Such additional work can involve producing more computer output and providing more extensive interpretation than the teacher asked for.

To determine what additional work you could submit, ask yourself what further information could be obtained with the computer that would be related to the assigned project and would be both relevant and interesting to you. A key point is that you make sure that any of your additional serious effort includes evidence of additional meaningful thought. Your extra achievement is likely to both enrich your learning experience and improve your grade on the computer assignment.

## SUMMARY

### How to Make Effective Use of Your Time Between Classes—Homework and Beyond

Try to answer the question on the left before you look at the answer on the right.

|  |  |
|---|---|
| 1. When should I do homework? | As soon as possible after the class, and definitely before the next class. |
| 2. When should I check my homework? | Preferably, as soon as I do it; definitely before the next class. |
| 3. How can I get a head start in preparing for a test? | I can do this by constructing a list of potential topics for the test as the unit is going on and not waiting for the period just before the test. |
| 4. How can I reinforce my knowledge of previously covered concepts during homework sessions? | Review concepts from my previous classes by reworking homework problems or working out new problems from those previous sections in my textbook. |

| | |
|---|---|
| 5. What other reading material can I use in the course? | Use study guides and review books as resources to help me in the course. |
| 6. How can I benefit from working with other students in the course? | Get the class notes and homework assignment if I miss a class. Check my solutions to homework problems with them. Encourage other students to ask me questions about the course material. |
| 7. What do I do if I have a question from homework problems or a previous class? | Try to get an answer to my question as soon as possible. |
| 8. What do I do if I don't naturally have a question? | Create questions about notes and homework to ask myself and then answer. |
| 9. What should I do with the few minutes before class? | If possible, use the few minutes just before class to review homework, class notes, or relevant textbook material. |
| 10. How can I break up my study sessions to maximize my learning? | Study in forty-five to sixty minute segments, with breaks that last between five and ten minutes. |
| 11. How can I do well on computer assignments? | I can start working on the assignment as soon as it is given. I can make sure I understand all the commands I use, and make careful, dedicated efforts to understand and interpret my output. |

## PERSONAL WORKSHEET

1. Here are three or more weaknesses that I have had when studying for a math class.

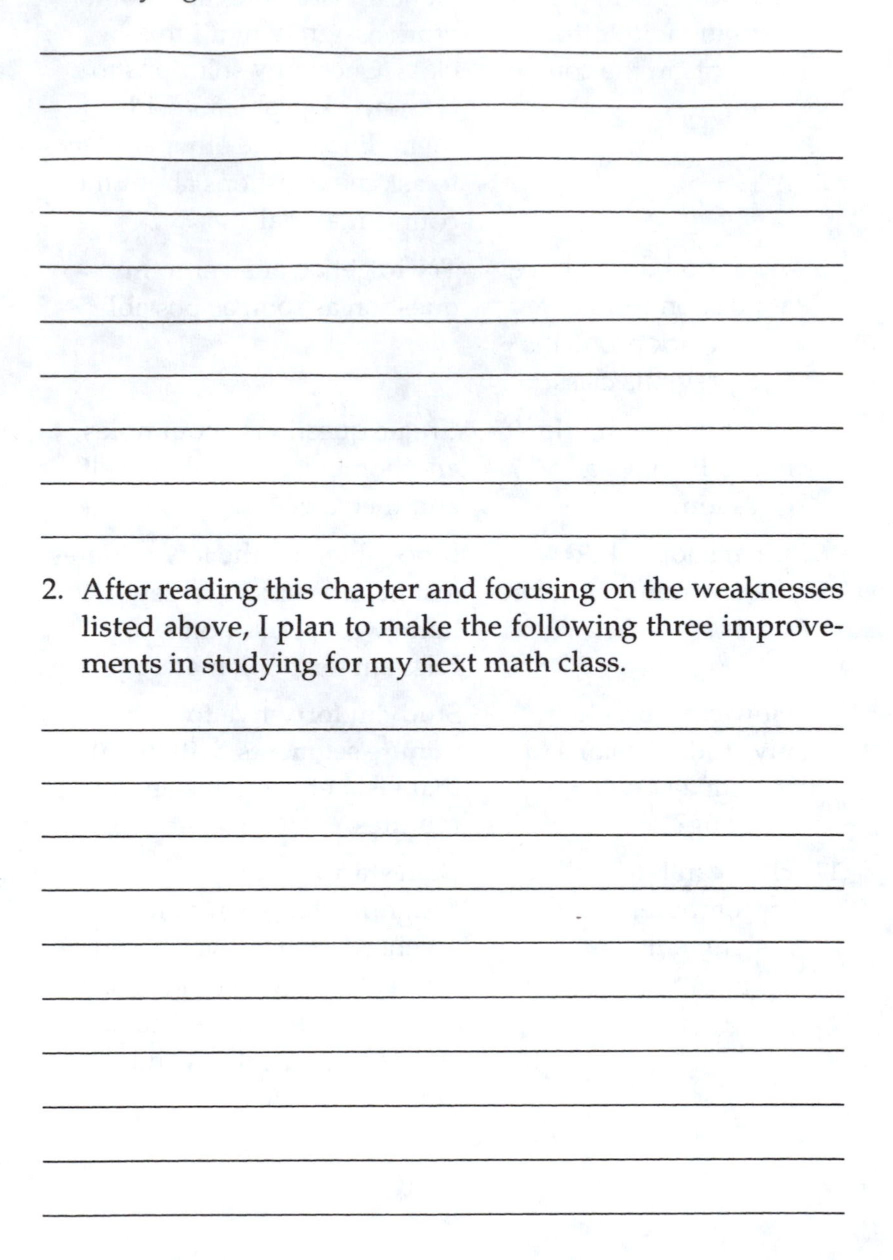

2. After reading this chapter and focusing on the weaknesses listed above, I plan to make the following three improvements in studying for my next math class.

## PRACTICE

If you are currently in a math course, go to a bookstore and find study guides or workbooks that deal with the subject of your course. Examine sections of the books that deal with topics you have already covered in class. Note the books that are most readable and understandable. Buy the book (or books) that are more readable and that contain the most practice problems and questions with solutions.

# Make Preparing for Tests a Sure Thing

# CHAPTER 7

# How to Aim for Perfection in Your Test Preparation

## APPLY THE NUMBER ONE GOAL OF TEST PREPARATION

The number one goal of test preparation can be stated simply:

**Always aim for 100%.**

That is, never seek merely to pass a test or to get a 70% or 80%. Always go for 100%. By focusing on this goal, you may be able to avoid ever again having a "mental block" during a test.

It is easier to focus on the goal of obtaining 100% than it is to focus on achieving a lower percentage, say 70% or 80%. If you aim for 70% or 80%, you have to make the difficult decision of evaluating how much studying is sufficient for you to achieve your goal. However, if you aim for 100%, you know that your goal is to master *all possible topics*. With such a clear goal, it becomes easier to establish a system for implementing it.

To strive for 100% on a test requires that you not allow yourself to have any weaknesses among the potential test topics. Out of laziness, carelessness, or misjudgment, you might feel that it is acceptable to be weak on one or two of the possible topics, so long as you think you have mastered *most* of them.

However, this strategy is not likely to work, because even if you allow yourself to be weak on only one topic, you are likely to worry during the test about that topic appearing. While approaching any given problem, you may very well be distracted by the possibility that knowledge of that weak topic will be tested somewhere in the problem. In this way, you run the risk of losing all confidence that you can succeed on the test. This loss of confidence can result in a nervousness that can make you unable to recall much of the knowledge that you do have. Students often label this lack of recall of knowledge during a test a "mental block." The topic of mental blocks is so important that it will be discussed in detail later in this chapter.

**You need to aim for 100% when you study for a test because:**

1. **The higher the grade you aim for on the test, the higher the grade you are likely to get.**
2. **Aiming for 100% helps ensure against having a "mental block" during the test.**

What should be your first step in aiming for 100%? The first thing you need to do is to make a list of all the possible topics that may appear on the test. When constructing this list, do not be tempted to omit a topic.

One of the study traps you may fall into that will cause you to omit topics is to try to guess what the teacher is or is not

going to put on the test. A typical comment about a potential topic that a student does not feel like studying very carefully might be:

"It probably won't be on the test."

If you could accurately guess which of the potential topics the teacher will not include on the test, then you could, of course, eliminate topics to study. However, it is impossible to read a teacher's mind. When you eliminate potential topics, you are, at best, making an educated guess as to which topics may not be included on the test. Since you cannot be certain you are correct in deciding which topics to eliminate, you are violating the number one goal of test preparation: Aim for 100%. This principle requires you to study all *possible* topics that may appear on the test.

Consider the case of Steven, who thought he had studied very hard for his college algebra test. He hadn't missed any classes. He worked out all the homework problems in the chapters on which he was to be tested. However, even with all his preparation, Steven still felt weak on one topic in one of the chapters. Since it was a topic that the teacher had discussed in class only briefly, Steven thought, "It probably won't be on the test, so I won't worry about it too much."

So Steven takes the test having a high degree of knowledge about all the topics except the one he neglected to master. During the test, he begins to get nervous, worrying that his omitted topic will appear on the test. At one point, he gives up on a problem whose solution is not immediately obvious to him, since he thinks that it might be a problem that requires knowledge of his omitted topic. In giving up on the problem, he becomes more nervous and insecure worrying about the points he is losing and fearing that he will not do well on the whole test. As a result of his nervousness, he begins to forget some of the material that he had mastered and to have trouble solving problems that he had been confident about just a few minutes before the test.

When his test is returned to him, Steven discovers that he got a 55. This was a test on which he thought he had studied well enough to get at least a 90. When the teacher asks Steven why he did so poorly, he says, "I really knew the material very well. I just had a mental block during the test."

## UNDERSTAND THE CAUSES OF MENTAL BLOCKS

The "mental block" is so pervasive a fear and concern of students in a math course that it deserves to be discussed in depth.

Students have mental blocks during a test when they cannot recall information and concepts that they thought they had mastered before the test. You are suffering from a mental block when all of the following conditions exist:

1. You study enough for a test that you believe you have at least some of the knowledge needed to take the test.

2. During the test, you discover that you are unable to answer questions about material you thought you knew before the test.
3. You have the feeling that previous knowledge has suddenly vanished from your mind during the test.

The surprise and suddenness of this loss of knowledge produces the feeling of mystery that dominates the concept of "mental block."

What causes a mental block? And how can you eliminate your chance of ever having one?

Remember Steven? Steven thought he "knew" the material before his test even though he omitted one topic from his studying. Suddenly, during the test, he was unable to recall much of what he thought he knew. In Steven's view, something happened to him during the test, something mysterious and beyond his control, which kept him from calling upon knowledge that he did, in fact, have before the test. Steven was wrong. He was not wrong about his having a mental block. He was wrong in thinking that the cause of his mental block originated *during* the test and that he had no control over it. Steven's mental block was caused by his lack of confidence resulting from his omitting that one topic from his studying.

Another possible cause of a mental block occurs when you have "studied" all the topics but by test time, you have only a "general idea" about some of them. This lack of a thorough knowledge of one or more topics will likely cause you to be unable to answer questions on these topics. This may result in the feeling that you have a mental block.

## AVOID MENTAL BLOCKS WITH PREPARATION

A mental block occurs during a test when there is inadequate preparation *before* the test. (In Steven's case, he neglected to master one of the topics that might appear on the test.) Such inadequate preparation tends to diminish students' confidence that they can succeed on the test. This loss of confidence can result in nervousness during the test, which can reduce their ability to recall much of the knowledge they had before the test.

The trick to avoiding mental blocks, then, is to build your confidence to a high level, based on a high level of knowledge. The way to do this is to prepare for the test so thoroughly and comprehensively that by test time you can honestly call yourself an expert on *all* the possible test topics. You should prepare so well that you *know* it is impossible for you to forget how to solve any problems or answer any questions on all eligible test topics.

**To eliminate your chance of ever having a mental block during a test, you need to achieve the highest quality of preparation before the test.**

## AVOID MAKING EXCUSES FOR NOT DOING WELL

If you don't do well on a test, you may think that you have a good excuse. You may blame your poor performance on an argument you had with a boyfriend or girlfriend just before the test, or perhaps on a headache, stomachache, or bad cold.

All of these excuses for not doing well are just that. Excuses. Offering such excuses to yourself or your teacher may make you feel better, but it only distracts you from the real cause of your failure. Regardless of the life or health problems that may concern you during the test, the real overriding reason that you didn't do well is that *you didn't prepare well enough.*

> **Solid preparation will allow you to do so well on a test that you won't need to use any of your "good" excuses.**

---

Here is a story concerning one of my own students that illustrates the point. One day as my students entered my classroom to take a test on the current unit, several of them warned me that one student, Mark, might not show up for the test. They told me that the previous night Mark had hit his head against the gymnasium wall during a soccer game. He now had at least a bad headache, if not something more serious.

Forewarned, I began administering the test to the class without Mark. About five minutes into the test, Mark walked into the classroom holding his head and looking sick.

At this point, I assumed that he was going to ask me when he could take a makeup test. But to my great surprise, he asked me if he could start the test right then. I asked him if he was sure that he was in good enough condition to take it. He said he wanted to take it then even with his headache, since he had studied well and wanted to get the test over with.

Not only did Mark take the test, he received a grade in the high 90s, the highest grade in the class. Later, I joked that perhaps I should send the rest of the class to hit their heads against that same wall.

If you have really prepared well for a test, you may not need to use a "legitimate" excuse for not doing well.

## BEGIN TO PREPARE FOR THE TEST EARLY

When should you begin your test preparation? You need to allow enough time to complete all the steps in the process.

**Begin your test preparation process at least a week before the test.**

You may doubt the usefulness of this advice, since the teacher is unlikely to have covered all of the potential topics for the test as early as a week before the test. However, most of the potential topics will have been covered, so that there should be plenty of material for you to begin the test preparation process.

## SUMMARY

### How to Aim for Perfection in Your Test Preparation

Try to answer the question on the left before you look at the answer on the right.

| Question | Answer |
|---|---|
| 1. What is the number one goal of test preparation? | Always aim for 100%. |
| 2. How do I begin my study process? | Construct a list of *all* possible topics that may appear on the test. |
| 3. What should I never say about a potential topic? | Never omit a potential topic from my list with the comment, "It probably won't be on the test." |
| 4. How can I eliminate my chance of having a mental block? | Eliminate the chance of having a mental block by thoroughly covering all the possible topics. |
| 5. When I make excuses for not doing well, what does that say about my test preparation? | It says that my problem is that I didn't study enough to do well. |
| 6. When should I begin my test preparation? | Begin my preparation at least a week before the test. |

## PERSONAL WORKSHEET

I. Check the statements that are true for you.

___1. I do not always work to get 100% on the test.

___2. I omit studying certain topics, thinking, "it probably won't be on the test."

___3. I sometimes have "mental blocks" during math tests.

___4. I sometimes overestimate the grade I will get on a math test.

___5. I sometimes make excuses for not doing well on a test.

___6. I sometimes wait until the last minute before starting to study for a test.

II. Check the top three changes that you plan to make in aiming for perfection in test preparation. Add any other strategies you plan to use.

___1. I will work to get 100% on the test.

___2. I will not omit any relevant topics in my studying for tests.

___3. I will work to achieve the highest possible preparation for the test.

___4. I will not make excuses for not doing well on a test.

___5. I will begin studying at least a week before the test.

**Other:** I will ____________________________________________

________________________________________________________

________________________________________________________

III. Rank the strategies you checked in section II according to how important you think they will be for your success. Begin by placing the number 1 to the left of the most important strategy.

IV. After you try these strategies, refer back to this page and highlight the ones that worked best for you.

## PRACTICE

If you are currently in a math course, use the back of your math notebook to begin to construct a list of topics to be covered on the next test. Use your homework assignments for the unit since the last test to help you construct your list. Next to each topic write the page and problem numbers of homework problems that deal with the topic. Make sure to keep adding to the list as the unit goes on.

# How to Make a List of Topics That Might Be Covered on the Test

## DETERMINE THE RULES AND SCOPE OF THE TEST

What is the first step you need to take when your math teacher announces a test? Before you even begin studying, you must:

> **Determine the rules and scope of the test.**

To do this, you have to obtain the answers to several questions.

What will the test cover? It is not enough to know which chapters in the textbook are included. You need to know if all topics in all chapters are included. Are there any topics, problem types, theorems, or definitions that the test will not cover? Are there any particular topics, concepts, or types of problems the teacher considers especially important? Ask if the test will involve solving problems, proving theorems, writing definitions, or answering multiple-choice, completion, or true-false questions.

Ask what, if anything, your teacher will allow you to bring with you to the exam room. For example, can you bring your textbook, notes, a list of formulas, or a calculator? It is important to realize that the rules may be different from test to test in any given course. For example, a teacher may allow you to use a calculator or a list of formulas for one test, but not for another. Make sure you are clear about the rules for *each* test.

## CONSTRUCT YOUR LIST OF TOPICS

Now that you have complete information from your teacher on the rules and scope of the test, what do you do next? You are now ready to begin the test preparation process. Your first step should be to:

> **Write out a list of all the topics that the test might cover.**

If you followed the advice in Chapter 6 on building a list of topics step by step as you work through a unit, you may not need additional information on this subject here. But let's assume you haven't done this (or, you would like additional advice on constructing a list of topics). How should you construct your list from scratch?

Your list of topics should be phrased as a series of actions to be performed. For example, a topic might be described in a sentence beginning with any of the following words or phrases:

"Calculate . . ."

"Find . . ."

"Evaluate . . ."

"Solve problems of the form . . ."

"Prove the result that . . ."

"Identify . . ."

"Simplify . . ."

"Distinguish between . . ."

When performing this task, it is important to:

**Make your list of topics as specific as possible.**

For example, suppose that for the test you need to know how to find the equation of the line passing through two points and, in addition, you need to know how to find the equation of a line passing through a given point parallel to a given line. Then your list of topics should read:

1. Find the equation of the line passing through two given points.
2. Find the equation of the line parallel to a given line passing through a given point.

Now you might, instead, just write these two topics as one:

1. Find the equation of a line.

However, the first alternative of writing the two more specific topics is far more useful.

As another example, consider "word" problems, which many students find difficult to solve. They have trouble extracting the needed information and determining how to use it to answer the question. Word problems need to be handled like any other topic on your list.

**List each kind of "word" problem separately on your topics list.**

Be as *specific* as possible. For example, here are two familiar kinds of problems:

1. A "work" problem:
   If Jack takes 3 days to do a job and Jill takes 5 days to do the same job, how long will it take both of them to do the job if they work together?
2. An "investment" problem:
   If Sarah invested $2000 at 6% interest, how much would she have to invest at 9% so that her total interest per year would equal 8% of the two investments?

If both types of problems were covered in your unit, your topics list might include the item:

1. Solve "word" problems.

A much more useful listing would be:

1. Solve "work" problems.
2. Solve "investment" problems.

This more specific breakdown will make it much easier for you to focus on each type of problem in your test preparation.

To make it easier to construct your list of topics, consider examining the table of contents in your textbook to find the section that deals with the relevant chapters. This would be most helpful if the sections of each chapter were relatively short; in that case, each chapter section might suggest a topic to be listed. However, if several topics were discussed in each section, the table of contents might be less helpful.

## FIND SPECIFIC PROBLEMS FOR EACH TOPIC ON YOUR LIST

For each topic on your list, make a list of problems from the book or study guide that deal with the topic. The list should be long enough so that if you could do all the problems, you would be certain of mastering the topic. For some topics, you may need only three or four problems to feel confident on the topic. Other more difficult topics or those with many subtopics may require ten to twenty problems or more.

Place each list of problems next to its topic. For example, suppose one of the topics involves solving linear equations with one unknown, and there are twenty questions on page 50 of your algebra book that deal with the topic. Perhaps the textbook has answers to all the odd-numbered problems, and the even-numbered problems require the same type of problem-solving skills as do the odd-numbered problems. Then, one way in which this topic may be listed in your topics list is as follows:

1. Solve linear equations with one unknown.
(50/1–20 odd)

The parentheses contain the page and problem numbers of a list of problems that deal with the given topic.

**When you construct your list of problems for each topic:**

1. **Make it long enough to provide sufficient practice for you to master the topic.**
2. **Make it complete enough to include all the types of problems on the topic that might appear on the test.**
3. **Make it diverse enough to include problems on both a moderate and a challenging level of difficulty.**

Repeat the process of writing the topic together with the problem page number and the item numbers in parentheses for each possible topic.

Here is a sample list of topics in an algebra course:

1. Solve linear equations in one variable. (57/1–13 odd)
2. Solve for one variable in terms of another. (57/15–19 odd)
3. Solve simultaneous equations. (57/21–35 odd)
4. Find the slope of a line through two given points. (78/1–19 odd)
5. Graph a linear function, given the equation. (108/1–15 odd)
6. Find the equation of the line through two given points. (109/33–51 odd)
7. Evaluate functions for particular values. (98/1–9 odd)

It should be emphasized that you must:

**Make sure to include all topics on your list that may appear on the test.**

If necessary, check with your teacher that you have not omitted any topics that *may* be covered on the test.

## SUMMARY

### How to Make a List of Topics That Might Be Covered on the Test

Try to answer the question on the left before you look at the answer on the right.

| | |
|---|---|
| 1. What's the first thing I should do when my teacher announces a test? | Ask my teacher to tell me the rules and scope of the test. |

| | |
|---|---|
| 2. How should I begin my test preparation process? | Make a list of topics that might be covered on the test. |
| 3. What should I include in a topics list for a test? | Include all possible topics that might be covered on the test. |
| 4. What should I include for each topic on my list? | Include a reference list of problems from my textbook or other sources that deal with the topic. |
| 5. How difficult should the problems be that are chosen for each topic? | In addition to the typical, standard level of problems, choose some of the more challenging problems. |

## PERSONAL WORKSHEET

I. Look at the following list of questions that you can ask your teacher about a test. Rank them according to how important you think getting the answer is for your own success. Begin by placing the number 1 to the left of the most important question.

___ 1. Can I use a calculator?

___ 2. Can I bring a list of formulas?

___ 3. What are all the topics that will be tested?

___ 4. What kinds of questions will there be: problems, multiple-choice, completion, and so on?

___ 5. How many questions will there be for each type of question?

II. Check the top three changes that you plan to use in constructing a list of topics that might be covered on the test. Add any other strategies you plan to use.

___ 1. I will include all possible topics in my written list.

___ 2. I will make my list of topics as specific as I can.

___ 3. For each topic, I will include all types of problems.

___ 4. For each topic, I will include enough problems to master the topic.

___ 5. For each topic, I will also include the more difficult problems.

**Other:** I will ______________________________

______________________________________________

______________________________________________

III. Rank the strategies you checked in section II according to how important you think they will be for your success. Begin by placing the number 1 to the left of the most important strategy.

IV. After you try these strategies, refer back to this page and highlight the ones that worked best for you.

## PRACTICE

If you are currently in a math course and have not yet constructed a list of topics to be covered on the next test (first discussed in Chapter 6), construct such a list now. Be sure to include the page number and problem numbers of homework problems that deal with the topic. If you have already constructed your list, try to improve the list by making topics more specific, if possible.

# How to Master Each Topic

## AVOID DECEIVING YOURSELF ABOUT HOW WELL YOU STUDY

> "I studied for my math test for 19 hours and I only got a 37 on it."

Have you ever made a statement like this? Do you know anyone who has? You may be tempted to make this kind of comment when you believe that you have worked hard preparing for a test but end up receiving a low grade.

If you receive a low grade on a test after putting in a lot of time studying, it usually indicates that you have used ineffective study methods to prepare for the test. That you made a statement such as the one above indicates not only that you have used ineffective study methods, but even worse, that you didn't realize that you had.

> **Most of your study time should be spent *writing out* and *thinking about* the concepts. Only a small percentage of study time before the text should be reading your notes and textbook.**

This chapter and the next will tell you specifically the kind of writing you need to do to prepare for a math test.

## MEMORIZE AS LITTLE AS POSSIBLE

Many students believe that math courses primarily involve formulas, definitions, and symbols, all of which need to be memorized. While taking the course, these students look for any opportunity they can get to memorize something. For these students, memorization is used as a crutch so they can avoid the effort needed to obtain an understanding of the concepts. Students may do this out of fear or laziness. They fear that they will never be able to understand the concepts, or they just do not want to do the work that it takes to master the concepts. Memorizing seems to be the easy way out. It is, however, the wrong way out.

**When you emphasize memorization in your test preparation, you are probably not thinking about the concepts. This can only lower your chance of success on the test.**

What's wrong with memorizing a formula you are expected to use, without making any attempt to understand it? If you do this, you are likely to have trouble knowing *when* to use the formula or whether an answer you get by substituting into the formula makes any sense. This approach is no way to "aim for 100%" on the test.

What's wrong with memorizing a procedure without making any attempt to understand it? If you do this, you will likely have more difficulty trying to use the procedure for a particular problem, or you will, at the very least, feel more insecure when trying to use it. If you don't *understand* a procedure, this is a weakness you need to eliminate before the test.

What is the best way to deal with formulas that you have to know for the test? You can memorize the formula if you need to, but it is critical that you also try to understand what it is saying. To do this, you should answer these questions about the formula:

1. When is the formula used?
2. Are there similar problems where this formula is *not* applied?
3. Are there any patterns in the formula that will help me remember it or how to use it?
4. How is this formula similar to or different from other formulas that may also be covered on the test?
5. How is the formula derived? (For some formulas, it may be helpful to learn how to derive them, even if your teacher doesn't require you do it.)

You should try to answer these questions even if your teacher will allow you to bring a list of formulas to the test or will provide you with such a list at the beginning of the test.

Suppose that on the test you will be expected to use some procedure that you have been taught. Instead of *memorizing* the procedure when studying for the test, try to answer questions such as the following:

1. Why is each step a part of the procedure?
2. Why are the steps applied in the order they are given?
3. How can I recognize what kinds of problems can be solved by using the procedure?

In many cases, you will find that after answering these questions about a formula or procedure, you will actually remember it without making any special effort to memorize it.

If you really need to memorize a formula, try saying it aloud each time you use it. It might then become fixed in your mind with little difficulty. Remember, though, that memorizing a list of formulas or steps in a procedure with no attempt at understanding gives you only a superficial knowledge of the concepts. Students who treat formulas and procedures in this way rarely do well on tests. In general, if you find yourself spending lots of time trying to memorize during your preparation, it may be a sign that you need to change strategies.

In addition to formulas and procedures, you may also need to memorize the statements of some definitions, theorems, and relevant symbols. Even in these categories, in addition to memorizing, make sure you can explain the meaning and use of the definitions, theorems, and symbols.

**Memorizing formulas and procedures without understanding them makes it more difficult to apply them. Formulas and procedures can be memorized, but, first and foremost, they should be understood.**

## WRITE SOLUTIONS TO PROBLEMS DEALING WITH EACH TOPIC

At this point, you have a *complete* list of all topics that you might be tested on, together with lists of problems that deal with each topic. The next step is to master each topic. You do this by working out the problems dealing with each topic.

> **It is critical that you work out problems *one topic at a time.***

For example, to prepare for his algebra test, Brian compiled a list of twenty topics with a list of problems from his textbook for each topic. Brian began to work out problems on topic 1. He worked out four of the fifteen problems listed for the topic. Since he felt that he now had a general idea of how to do the problems on that topic, he stopped dealing with that topic and went on to topic 2. After doing a few problems on that topic, he had a general idea of how they worked, and so he went on to topic number 3. He went through all twenty topics in this way, going on to a new topic after deciding that he had a general idea of how to do problems on the previous topic.

After Brian worked through all the topics, he decided to review by returning to topic number 1. He discovered to his dismay that he could not remember how to do these problems. He then discovered that he had the same difficulty with several of the other topics from the list.

Brian had trouble with those topics because he worked out too few problems dealing with them before going on to other topics. Although he did enough to acquire a general idea of how to do problems on these topics, his knowledge was just not deep enough to justify his confidence that he could work out such problems in the future. He never really mastered these topics because he didn't do enough before leaving them.

Brian had committed the common study error of not sticking to one topic at a time. The one-topic-at-a-time goal requires you to work out enough problems on a topic until you have developed *total confidence* in your ability to do problems on that topic. Such confidence requires doing more problems than you would do if you merely desired to achieve a general familiarity with the topic. Assuming you don't get stuck on any topic:

> **Do not begin dealing with the next topic until after you have *total confidence* on the current one.**

Sometimes, however, you may be stuck on a topic. For example, suppose you have achieved *total confidence* on topics 1, 2, and 3, but have reached a dead end on topic 4. What should you do? In this situation, you should *temporarily* skip topic 4. Continue to work through the problems that deal with your list of topics until you achieve *total confidence* on the rest of the list.

Make another attempt at topic 4 after you finish the others. Perhaps your work on later topics will help you handle topic 4. If you need to get help from review books, other students, or your teacher, do it. One way or another, master topic 4.

It is all right to temporarily omit topics that you have too much trouble with as long as you eventually achieve *total confidence* on these troublesome topics before the test.

If you stop working on a topic before reaching the *total confidence* level, you run the risk of having to face Brian's problem—that is, of later forgetting how to do problems on a topic you thought you had already mastered. If you don't do enough problems to master a topic before going to the next topic, you will likely have wasted all the time you spent working on the topic.

You might notice that there are some topics on your list that you are afraid of or dislike. You might find yourself hoping that these topics will not appear on the test. Your fear or dislike of a topic is an important clue that you find the topic difficult, that you are weak on the topic, and that you probably are not close to having total confidence on that topic. This is the kind of topic that you must make a *persistent* attempt to master. One clue that you have mastered a topic is if you get to the point that you *hope* it will appear on the test.

**If you are afraid of or dislike a topic, you must make a special, persistent attempt to master the topic in order to achieve total confidence on it.**

Mastering the topic will eliminate your fear and lessen your dislike of it. By the time of the test, you should not be afraid of or have a special dislike for any topic on your list.

## APPLY THE ACE STEPS

When students talk about planning to do very well on a test, they sometimes say that they plan to "ace" the test. The primary goal of this book is to help you learn how to ace math tests.

To ace a math test, you need to begin by achieving the one-topic-at-a-time-until-total-confidence goal. The primary procedure for accomplishing this goal is to complete the three "ACE steps" as follows. For each topic on your list of topics:

1. **Answer** each problem by working it out in writing. If you had to look at the answer or your notes or textbook before finishing the problem, write out the problem again from start to finish. It doesn't count until you have written the solution without looking at the answer, your notes, or your textbook.

   Don't limit yourself to working out only the easiest problems on a topic. Try some of the more challenging problems that the teacher may cover on the test.

**C** 2. **Confirm** that each answer is correct after you work out the problem. If it is not, you need to begin the problem again.

**E** 3. **Examine** your understanding of each problem after you have confirmed that your answer is correct. Do this by asking and then answering the following questions:

"What was the point of this problem?"

"What common errors do I need to avoid when working out problems of this kind?"

Apply the ACE steps of Answer, Confirm, and Examine to *every* problem you do on the first topic. Complete this procedure before you change topics. In this way, you can achieve total confidence on that topic. After applying the ACE steps to each problem on the first topic, repeat the process with the problems on the second topic, then the third, and so on, until you have total confidence on every topic on your list.

The advice in the Answer step—to practice the more demanding problems—needs to be highlighted:

> **In order to be able to work out very difficult problems on a test, you have to practice working out these kinds of problems ahead of time.**

The Confirm step requires finding some way of verifying or checking that your answer is correct. It would be helpful for you to apply the suggestions in the Chapter 6 section, "Check Your Homework Before the Next Class" on pages 87–91.

## COMPLETE THE REVIEW

After finishing the ACE steps for all the topics, you have to complete one final step to achieve the one-topic-at-a-time-until-total-confidence goal. This is the Review step.

The Review step requires that you reread your notes and relevant textbook sections dealing with the topics on your list. Pay particular attention to those questions you wrote in the margins and backs of pages of your notebook and in the margins of your textbook. As you read, place a question mark next to any item in your notes or textbook that you have the slightest question about. By the time of the test, you must get *all* of these questions cleared up, either on your own or with the help of your teacher or knowledgeable students.

**To achieve the one-topic-at-a-time-until-total-confidence goal, you must:**

1. **Complete the ACE steps of Answer, Confirm, and Examine for every problem you do for each topic on your list.**
2. **Take the Review step of clearing up all possible uncertainties in your notes and textbook involving the topics on your list.**

Achieving total confidence on all topics naturally improves your chances of doing very well on the test, since it greatly decreases your chances of being exposed to a test problem of a type you cannot solve. In addition, as discussed in Chapter 7, this procedure greatly diminishes your chances of having a mental block during the test.

If you want to be exceptionally dedicated, you might consider applying the ACE steps earlier in the unit as you are building your list of topics. After the teacher has covered, say, three topics in the unit (so that there are three topics on your list at that point), apply the ACE steps to those three topics at that time, *without waiting until you have completed the entire unit.* After mastering each of the three topics, one at a time, try selecting problems from the three topics at random and then working them out. After that, every time a topic or two is added to your list, again apply the ACE steps to all listed topics, and again work out problems at random from the updated list. If you continue in this way throughout the unit, you will likely acquire such a deep level of understanding and knowledge of the unit that you may need hardly any additional work before the test.

## SUMMARY

### How to Master Each Topic

Try to answer the question on the left before you look at the answer on the right.

1. What strategy will allow me to achieve mastery of all the possible topics?

   I can most easily master all the topics by achieving mastery one topic at a time.

2. When should I not attempt to memorize in my studying?

   I should never attempt to memorize a formula (or rule, procedure, or proof of a theorem) until after I have made a significant attempt to understand it.

3. What is the goal of the ACE steps and Review procedure?

   The ACE steps and Review procedure is intended to help me achieve the one-topic-at-a-time-until-total-confidence goal.

4. What are the three ACE steps?

   For each topic:

   1. *Answer* each problem by working it out in writing.
   2. *Confirm* that the answer is correct.
   3. *Examine* my understanding of the problem by asking myself the following questions:

"What was the point of the problem?"

"What errors do I need to avoid when working out problems of this kind?"

5. What is the Review procedure?

This is the final step to achieving the one-topic-at-a-time-until-total-confidence goal. It requires that I clear up all of the questions and uncertainties in my notes and textbook concerning the topics on my list.

## PERSONAL WORKSHEET

I. Check the top five changes that you plan to use in mastering your list of topics that might be covered on the test. Add any other strategies you plan to use.

___ 1. I will spend most of my study time before a test writing out and thinking about the concepts.

___ 2. I will focus more on understanding formulas and procedures than on memorizing them.

___ 3. I will memorize formulas and procedures only as a last resort.

___ 4. I will begin to work on the next topic on my list only after I achieve *total confidence* on the current one.

___ 5. If I fear or dislike a topic, I will make a special effort to master it.

___ 6. Before the test, I will practice working on the more challenging problems on a topic.

___ 7. I will complete the ACE steps of Answer, Confirm, and Examine for every problem I do for all topics on my list.

___ 8. I will review by clearing up all weaknesses on topics from my notes and textbook.

**Other**: *I will* ______________________________

______________________________________________

______________________________________________

II. Rank the strategies you checked in section I above according to how important you think they will be for your success. Begin by placing the number 1 to the left of the most important strategy.

III. After you try these strategies, refer back to this page and highlight the ones that worked best for you.

## PRACTICE

1. Assuming you have constructed a topics list as suggested in the Practice section at the end of Chapter 8, apply the ACE steps of Answer, Confirm, and Examine for every problem you do for all topics on your list.
2. Write down three formulas, definitions, or theorems that you have recently memorized in a math class. For each of these, check your understanding by:
    a. writing down an explanation and describing in what situations it applies.
    b. writing down an example.
    c. explaining it to another student without looking at your notes.

CHAPTER 10

# How to Be a Perfectionist When Preparing for a Test

## DISTINGUISH ONE TYPE OF PROBLEM FROM ANOTHER

Achieving the one-topic-at-a-time-until-total-confidence goal is often not enough to guarantee that you will do extremely well on the test.

On some tests, you may have an additional study challenge arising from the fact that problems on a test may be placed in random order. The first challenge of such a test is to identify the type of problem you are faced with before you attempt to solve it.

To do this, special preparation procedures are needed. Note, however, that the following procedures should be used only *after* the ACE steps and Review procedure of Chapter 9 have been applied for all the topics.

**Math tests in which problems seem more difficult because they are placed in random order require separate additional preparation procedures.**

1. Think of ways to distinguish each type of problem from any other that may appear on the test. Visualize in your mind how they are similar and how they are different. Write down a list of the similarities and differences.
2. Obtain or make up sample tests that include comprehensive collections of all the types of problems that may appear on the test. Some sources of these problems include:
   - sample tests from the teacher or other students
   - review exercises in your textbook
   - competing textbooks, or study guides purchased in bookstores

   Examine the problems on your list one at a time. For each problem, describe to yourself the correct method for solving that problem and why that method is correct. Do not work out each problem as you go along, since at this point you are only exercising your ability to determine the *method* for solving the problem.
3. Check that you have accurately identified the correct method for solving each problem.
4. Write out solutions to all of these problems. After doing a particular problem, state to yourself how you decided on the correct approach. Don't forget to check your solutions by applying the

suggestions in the Chapter 6 section, "Check Your Homework Before the Next Class" on pages 87–91. You can also verify your solutions either from a collection of solutions or with the help of other students or the teacher.

If you have collected one or two tests that your teacher has given previously, you may be tempted to work hard to master these tests without first making the effort to master all the topics on your topics list. This strategy is very risky. If you work solely from the previous tests without doing the necessary preliminary work for the current test, you might memorize how to do only the *particular* problems that appeared on *those* tests. Then, if the teacher includes problems and questions on your test that are even a little bit different from those on the previous test, you may be lost.

**To guarantee success on your math test, you must master all the topics on your topics list *before* you work on any practice tests.**

## FOLLOW THE "DRESS REHEARSAL" PRINCIPLE

Now you have applied the ACE steps and Review procedure to your topics list. Furthermore, you have determined that you can solve problems when they are presented to you in random order. At this point, are there additional effective study techniques you can use before taking your test? Yes! You can follow the "dress rehearsal" principle of test preparation.

**The best way to ensure success on a test is to take and master a "practice test" that has the same form as the actual test that you are preparing to take.**

It is vital that this practice test be at least as difficult and comprehensive as the test you are expecting to take.

To do well on the real test, it is beneficial to come as close as possible to experiencing it before actually taking it. To use the "dress rehearsal" principle of test preparation, follow these steps:

1. Master each topic first. (See Chapter 9.)
2. After finishing a unit or chapter, ask your teacher: "What kinds of questions or problems should we be able to deal with? What important questions, issues, or examples, do you think we should be looking at in these chapters?"
3. Ask your teacher what the form of your test will be. Specifically, determine from your teacher the kinds of problems, proofs, and short answers that will appear on the test. Will the test contain only problems? Some problems and some short-answer questions? If there will be short-answer questions, will they be multiple-choice, true-false, or completion? Other types?
4. Find old tests on the assigned material that your teacher or other teachers have given in the past. Obtain as many of these as you can find by asking other students.
5. Learn how to distinguish one type of problem from another by practicing the suggestions in the last section.
6. Make up your own "practice test." You can do this using exercises from textbook sections, review problems at the end of chapters in the textbook, questions and problems from available study guides and course outline books. In addition, you can have a friend, in or out of your class, select problems for you. From these resources, construct a collection of problems, questions, and short-answer questions of the types

that you have determined from your teacher are likely to appear on the test.

It is critical that the questions you select provide a *comprehensive* collection of what is likely to appear on the test. That means that the practice test must include questions on *all* possible topics on your topics list.

Many teachers select their test questions from the even-numbered questions in the textbook (since the odd-numbered questions were assigned for homework) and from any study guide that goes with the textbook. Therefore, if your teacher assigns the odd-numbered problems, include at least some of the even-numbered problems in any practice test that you create. Remember that you need to be able to check your answers with the help of the teacher or other students, if necessary. If you have found a study guide for your course, work out as many of the exercises as you have time for on the relevant topics in the guide.

7. After finding or constructing your practice test, take this test as a test by *writing out* your answers to all the questions you have collected. Do not look at solutions or refer to your notes or textbook while you are taking practice tests. You might decide to give yourself a time limit if you can decide what a reasonable time limit would be. If you do not have correct solutions ahead of time, you can check your answers by asking your teacher or other students in the class to verify your solutions. You should not be satisfied until

you have written correct solutions to all the questions on your practice test.

8. At the end of this process, make a list of any weaknesses you still might have among the possible test topics. If you still have any, work on those topics in any way you can, with or without help, until they are no longer weaknesses.

**Only when you have no weaknesses are you ready for the test.**

When you have finished this process, you will be a lot less likely to be surprised when you take the actual test. Since you already will have challenged yourself with a list of questions at least as difficult as those on the actual test, you already will have identified and eliminated any areas of weakness that you may have had on test issues.

**Only when you have taken and done well on a comprehensive practice test can you achieve maximum confidence that you will do well on the actual test.**

## BECOME FAMILIAR WITH THE OVERALL TEST PREPARATION PROCEDURE

The test preparation process involves the following steps, which have been discussed in Chapters 7–10:

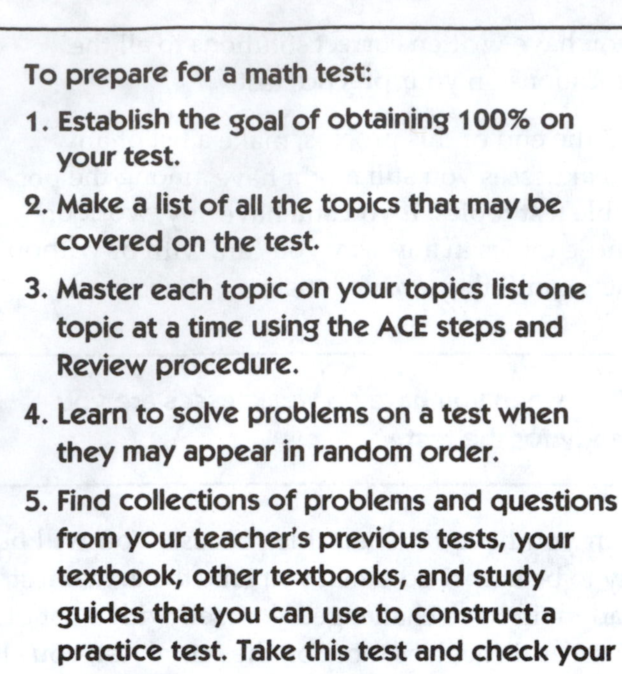
**To prepare for a math test:**

1. **Establish the goal of obtaining 100% on your test.**
2. **Make a list of all the topics that may be covered on the test.**
3. **Master each topic on your topics list one topic at a time using the ACE steps and Review procedure.**
4. **Learn to solve problems on a test when they may appear in random order.**
5. **Find collections of problems and questions from your teacher's previous tests, your textbook, other textbooks, and study guides that you can use to construct a practice test. Take this test and check your solutions.**

## STUDY EFFECTIVELY FOR A CUMULATIVE FINAL EXAM

Of all math tests, cumulative final examinations are often the most intimidating to students. On a cumulative test at the end of the term, you are required to know several times more material than you did for any individual test during the term. In addition, if you did not do well on any one or more of the earlier tests, you may, again, be responsible for the same material that you failed to master previously.

However, if you are organized and leave enough study time, you can be as successful on a cumulative final as on a smaller test. Use the system of test preparation already described along with these additional steps for dealing with a cumulative final examination in math:

1. Begin studying for your final one to two weeks before the exam.

2. Obtain as much information as possible from your teacher about what the test will cover.
3. Gather all the tests you have taken in the course as well as your topics lists for each test.
4. Verify that you can do problems and answer questions on all the topics on all your lists. For many of these topics, this will be easy, since you will be reviewing material you have learned fairly well already. Pay special attention to *all* topics that you feel weak on.
5. Study your previous tests enough so that you can answer similar but different questions. Pay special attention to any questions that you lost points on, as well as to questions you didn't lose points on but feel weak on anyway.
6. Work out review problems or practice tests at the end of the relevant chapters.
7. Obtain a sample cumulative final exam for your course or make one up using the system for creating practice tests provided in this chapter. Practice taking this exam.

## OVERSTUDY FOR THE TEST

After receiving a low test grade, students sometimes say that perhaps they studied too much.

> **It is not possible to study too much. It is not possible that overstudying can actually lower your grade. Even after you attain great mastery of the material, doing more work can only help you.**

Suppose that you have followed the advice given in this book. You have seriously studied toward the goal of attaining a 100% on your math test. You have achieved an extremely high level of knowledge on all the possible topics. You have worked with other students. You are confident that you can answer questions from other students on all the topics. You believe that you have *no weaknesses* that still need to be corrected.

Even in this case, if you find yourself with extra time, you can still do more. You can always find more problems to work out. You can always find another student to whom you can offer to explain concepts. You can always make up a new practice test. There is no limit to what you can do. And if you find that doing more work increases your confusion, you really didn't understand the material in the first place.

Let's look at a graph that shows the relationship between the hours a student studies for a test and the estimated test grade (in %) the student can expect.

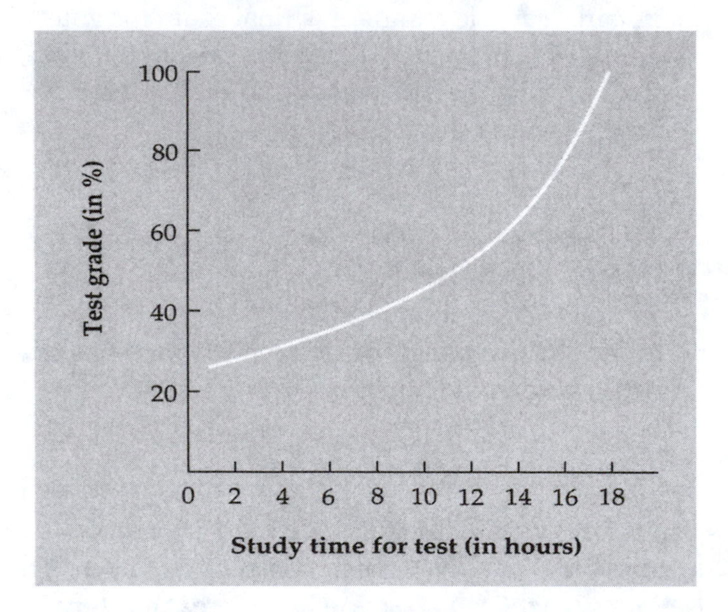


Let's suppose that it would take you 18 hours of effective study time to achieve 100% or close to 100%, as indicated on the graph. (This would require you to eliminate all your weaknesses.) On the left side, the graph shows that if you do not study at all for a test you probably would still receive some points. The first few hours of studying do not add very many points to your expected grade. But as your study time increases, the *rise* in your expected grade increases. Now look at the sharp rise that occurs near the end of what your study time ought to be if you are effectively aiming for 100%. Suppose that after you have studied for 16 hours, you look back and comfort yourself by focusing on those 16 hours instead of on the fact that you still need 2 more hours of work. The graph suggests that you should not expect more than approximately 75% or 80% on the test. It is the last two hours of study that lead to the greatest increase in your grade percentage.

The point of the graph is that:

> **If you are effectively aiming for 100% on a test, it is those last two or three hours of your study time that give you the greatest number of points. These are the hours you should be spending eliminating your last few weaknesses.**

One indicator of how you will do on a test is how you feel just before the test is passed out. If you walk into your test with any of the following feelings, it is a bad sign:

"Well, at least I think I'll pass."

"I think I'll get a 70%."

"I think I'll get an 80%."

"I have a general idea of how to do most of the problems."

"I know most of it, but I'm kind of weak on two or three topics."

Here's how you can expect to feel if you've taken the correct approach:

**If you have studied correctly, when you walk into the test you should honestly feel that:**

1. **you have no weaknesses among the eligible topics.**
2. **you have a reasonable chance to get 100% or close to 100%.**

## USE THE "SELF-CHECK OF STUDY SUCCESS" TO PREDICT HOW WELL YOU WILL DO ON A MATH TEST

Wouldn't it be nice if you could accurately predict how well you will do on a test before you take it? This chapter gives you a very good way of doing just that.

The following survey, called the "Self-Check of Study Success" (SCSS), is intended to help you evaluate how prepared you are for your next math test. Complete the survey at least once several days before your test. Identify any questions on which you scored less than 4. How you answer these questions will provide you with a clear idea of how much more you need to study before taking the test.

To calculate your total score, add up the numbers you circled for each question. A minimum total of approximately 81 points is necessary before you are ready to take the test with the confidence that you can achieve an A on it.

**Honest answers of at least 4 to all questions and a minimum total of about 81 points on the SCSS are necessary before you are ready to take a math test.**

For example, if your answer to question 7 ("I understand every topic that is likely to be covered on the test"), was less than 5, you are indicating that you sense you have a weakness on this issue. Your response should be to examine your complete list of topics to locate the ones that you still feel weak on. Now go back to Chapter 9 and apply the ACE steps and Review procedure to upgrade all your weaknesses to strengths. Just before your test, you can take the SCSS one final time.

## COMPLETING THE SURVEY

Circle the number that most accurately represents your response to each of the following questions about your upcoming test.

## SELF-CHECK OF STUDY SUCCESS (SCSS)

| | *strongly disagree* | | | | *strongly agree* |
|---|---|---|---|---|---|
| 1. I have obtained or have made a list of *all* the topics that may appear on the test. Each topic in the list includes a collection of exercises that deal with the topic. | 1 | 2 | 3 | 4 | 5 |
| 2. I have worked on each topic until I have mastered it using the ACE (Answer, Confirm, and Examine) steps, and only then have I gone on to the next topic. I have done this for every topic on the list. | 1 | 2 | 3 | 4 | 5 |
| 3. I have studied well enough, not just to pass, but to get at least 90%. | 1 | 2 | 3 | 4 | 5 |
| 4. I have covered everything so well that I think I can get close to 100%. | 1 | 2 | 3 | 4 | 5 |

| | *strongly disagree* | | | | *strongly agree* |
|---|---|---|---|---|---|
| 5. I have studied all potential topics that I might be tested on, even if I believe that the teacher is unlikely to include all such topics on the test. | 1 | 2 | 3 | 4 | 5 |
| 6. For all the types of potential problems, I can describe the steps involved in solving them. | 1 | 2 | 3 | 4 | 5 |
| 7. I understand every topic that is likely to be covered be on the test. | 1 | 2 | 3 | 4 | 5 |
| 8. Within the past few days, I have correctly solved enough problems dealing with each possible topic that I previously had doubts about. I have done this without looking at the solutions before solving the problem. | 1 | 2 | 3 | 4 | 5 |
| 9. (If applicable) I have specifically practiced identifying the type of problem I am faced with, even when problems are given in random order. I feel confident that I can make such identifications. | 1 | 2 | 3 | 4 | 5 |
| 10. I have reached the point where I am sure I can work out all the problems on a sample test (if one is available) nearly perfectly without looking at the solutions. | 1 | 2 | 3 | 4 | 5 |
| 11. I can explain to another student how to solve *all* of the types of problems that may be on the test. | 1 | 2 | 3 | 4 | 5 |

| | *strongly disagree* | | | | *strongly agree* |
|---|---|---|---|---|---|
| 12. I can write the solutions to problems dealing with *all* the topics that may be on the test. | 1 | 2 | 3 | 4 | 5 |
| 13. I can solve all types of potential problems even when they are given in random order. | 1 | 2 | 3 | 4 | 5 |
| 14. For each type of potential problem, I can describe the typical errors a student might make in solving such a problem. | 1 | 2 | 3 | 4 | 5 |
| 15. Even though I attend class regularly, take complete notes, and do my homework, I have made an additional special effort to study for the test. | 1 | 2 | 3 | 4 | 5 |
| 16. I have obtained a collection of problems and questions that can be thought of as a comprehensive practice test. I have correctly worked out all problems on this practice test without looking at the solutions. I have done this only after mastering all the topics individually. | 1 | 2 | 3 | 4 | 5 |
| 17. I believe that I can answer most questions a student might have about issues that might arise on the test. | 1 | 2 | 3 | 4 | 5 |
| 18. I know the material so well that I think that I am going to enjoy taking this test. | 1 | 2 | 3 | 4 | 5 |

## SUMMARY

### How to Be a Perfectionist When Preparing for a Test

Try to answer the question on the left before you look at the answer on the right.

| Question | Answer |
|---|---|
| 1. What additional challenge arises on some tests when problems are placed in random order? | I may have to identify the type of problem I am faced with before I can attempt to solve it. |
| 2. What is required for such a test? | I must implement *special* additional preparation procedures. |
| 3. What special preparation procedures are required? | I need to write down the ways in which one type of problem may be distinguished from another. Follow this procedure:<br>1. Obtain sample tests or collections of problems of all the types that may appear on the test. Examine the problems on the list one at a time. For each problem, describe the correct method of solving that problem and why that method is correct.<br>2. Check that I have correctly identified the correct method for solving each problem.<br>3. Write out solutions to all of these problems. After doing a particular problem, I should state how I decided on the correct approach to each problem. Find a way to check my solutions. |

| | |
|---|---|
| 4. What must I do before I work on any practice tests to guarantee success in the course? | I must master all the topics on my topics list. |
| 5. What is the "dress rehearsal" principle of test preparation? | The "dress rehearsal" principle of test preparation states that the best way to assure success on a test is to take and master a practice test that has the same form as the actual test I am preparing to take. |
| 6. What must I do before creating a practice test? | Master each topic on my list, one topic at a time, and determine that I can solve problems when they are presented in random order. |
| 7. What are the important qualities of a practice test? | It must be at least as difficult and comprehensive as the actual test is expected to be. |
| 8. What should I do with my practice test? | Take the practice test by writing out answers to all the questions I have collected. |
| 9. What should I do after I have written out my answers to the practice test? | Find a way to check the solutions to the questions on the practice test. |
| 10. What special study procedure should I use for a cumulative final exam? | In addition to the system of test preparation described above, I should do the following:<br>1. Begin studying one to two weeks before the exam.<br>2. Obtain as much information as I can about what the test will cover. |

3. Gather all tests I have taken in the course and verify that I can do all problems.
4. Verify that I can do problems and answer questions on all the topics on all my lists.
5. Study previous tests so that I can solve similar but different problems.
6. Work out review problems or practice tests at the end of the relevant chapters.
7. Practice taking a sample cumulative final exam.

11. How can I predict my grade on a test?

I can predict my grade by taking the Self-Check of Study Success (SCSS).

## PERSONAL WORKSHEET

I. Rank the following strategies according to how important you think they will be to your success. Begin by placing the number 1 to the left of the most important strategy.

___ 1. I will master all the topics on my topics list before I work on any practice tests.

___ 2. I will make a special effort to learn to solve problems on a test when they may appear in random order.

___ 3. I will create and take a practice test that has the same form as the actual test that I am preparing to take. I will check my solutions to the practice test.

___ 4. I will spend the last few hours of my test preparation eliminating my last few weaknesses.

___ 5. I plan to walk into every test feeling that I have no weaknesses among the eligible topics.

___ 6. I will use the Self Check of Study Success (SCSS) before a test to evaluate how prepared I am for the test.

II. After you try these strategies, refer back to this page and highlight the ones that worked best for you.

## PRACTICE

1. Use the list of topics you constructed in the Personal Worksheet at the end of Chapter 6 or Chapter 8, as well as problems from your textbook, other textbooks, or study guides to create a practice test that deals with all the topics on your list.
2. About three days before your next math test, take the Self-Check of Study Success (SCSS). As you continue to study

during those last three days, take the SCSS at least twice more before the test to see if you are getting your confidence up to its highest possible level.

3. An hour before the test, predict your score on the test. Can you honestly predict that 95% to 100% is possible?

# CHAPTER 11

# How to Take a Math Test

It's the day of the test. You have concluded all of your test preparation. Is there anything more you can do to attain the highest possible grade? The answer is an emphatic yes!

## ARRIVE EARLY FOR THE TEST

Make sure that you arrive at the classroom for your test at least a few minutes before it starts. Leave yourself enough time to put your books down, settle into your seat, get out your writing materials, formula sheet (if allowed), and calculator (if allowed).

> **Arrive at the classroom early for your math test.**

## USE YOUR TIME SENSIBLY

Many tests do not allow the use of formula sheets. This makes some students nervous. They worry that under the pressure of the test they won't remember important formulas. If you feel this way, you may find it helpful to jot down these formulas somewhere on your test as soon as you receive it.

Before you actually begin working on the test, read the directions carefully. Then look through the entire test quickly to get an idea of the length of the test and the types of questions it contains. Taking the time limit and point values of questions into account, determine how much time you can allot for each question. If possible, leave a few minutes at the end to review your answers.

Unless a question is extremely simple, read it over two or three times. If you have prepared yourself in the way suggested, most of the questions on the test should not be too surprising. Don't read too much into them. Don't assume that a test problem is a trick problem—most test questions are not trick problems.

To build your confidence, you might choose to answer the easiest questions first. If you get stuck on a problem, you may get a clue by comparing it to similar problems you may have done in class or for homework. Whatever you do, don't spend so much time on the problem that you cannot attempt all the rest of the questions.

## USE SPECIFIC TEST-TAKING STRATEGIES

What should you do if you are unsure of whether to change your first answer to a question? The traditional advice was that your first guess was probably correct. However, if you have really studied effectively:

**Rethinking an answer that you are uncertain about might allow you to come up with a better choice. Don't automatically assume that your first answer is correct.**

Suppose you are attempting to solve a problem that requires you to use the answer to an earlier problem that you were unable to solve. Make up an answer to the earlier problem and use it at the appropriate place in the later problem. Most teachers will give you credit for a problem that you solve with correct logic, even if you obtain a wrong answer because of a wrong answer to an earlier problem.

Show your work for all the questions you answer, especially if you know that the teacher's policy is to give partial credit to a student who shows some correct work on a problem without getting a correct final answer.

Make sure to do at least some work on every question. Never leave an answer blank. Writing down something you know that is relevant to the question gives you a chance of getting it correct or at least receiving some partial credit.

Here are some specific problem-solving strategies that you can use when you don't know how to begin your work on a problem:

1. Write down the information you are given.
2. Write down what you want to find or prove.
3. Write down any formulas, definitions, or theorems that may help.
4. Try to use all the relevant information you are given.
5. Try to get a clue to the problem by estimating what the answer will be. Write that down.

6. Try to work backward from the conclusion to get a hint of what the process might be.
7. Compare any final answer you get with your estimate to see if the answer makes sense.

If you finish the test early, check your answers as carefully as you can. Don't be in a rush to leave the room, or you will lose a valuable opportunity to catch any errors you might have made.

If you have time, try to check some of your answers by applying the suggestions in the Chapter 6 section "Check Your Homework Before the Next Class" on pages 87–91.

## MAKE EFFECTIVE USE OF THE TEST WHEN YOU GET IT BACK

When your test is returned, check it to see that it was graded correctly. When the teacher goes over the test in class, make sure to locate your errors. Write corrections on your test in a different color pen or pencil to distinguish it from your original answer. If you are still puzzled as to why your answer was incorrect, ask your teacher for clarification. One way or the other, clear up every question you have about the test. Make sure not to erase your earlier mistakes! Leaving them there will encourage you to focus on them so that you can more easily avoid making them again on a future test.

**Always use graded tests to help you identify ways to improve your study habits for the next test.**

Determine the deficiencies in your study habits that caused you to make an error or be unable to answer a question on the test. For example, when you studied for the test, did you:

- include all possible topics?
- cover all topics thoroughly?

- prepare yourself to identify the different types of problems that appeared on the test?
- work effectively with practice tests?

It may be tempting to blame your teacher or the test for an unsatisfactory grade, but you will do better to identify improvements you can make in your study habits. If you can accurately identify what you did wrong in your studying, you are less likely to repeat these test preparation errors on future tests.

## SUMMARY

### How to Take a Math Test

Try to answer the question on the left before you look at the answer on the right.

|  |  |
|---|---|
| 1. When should I arrive for a test? | Arrive early at the classroom. |
| 2. What should I do as soon as I receive the test? | Read the directions carefully. Look through the test quickly to estimate how much time to allot each question. |
| 3. Which questions should I answer first? | Answer the easiest questions first. |
| 4. What should I do if I am unsure whether to change my first answer to a question? | Rethink the answer. Don't automatically think the first answer is correct. |
| 5. What should I do if I finish the test early? | Check my answers. Do not rush to leave the room. |
| 6. How can I use my graded test when I get it back? | Use it to help identify ways to improve my study habits for the next test. |

## PERSONAL WORKSHEET

I. Circle the letter of the answer that is true for you.

1. I usually arrive ___ for a test.
   a. early
   b. exactly on time
2. When I receive the test, I begin working on it ___.
   a. after I look through the entire test quickly
   b. immediately
3. I try to solve ___.
   a. all the problems
   b. only the problems I am sure of
4. If I am unsure of an answer, I ___
   a. rework it
   b. leave my first answer
5. I ___ show my work for a problem.
   a. always
   b. don't always
6. When I have finished the test, I ___.
   a. check all my answers
   b. leave immediately
7. When I get my test back, I ___ my mistakes.
   a. correct
   b. erase

II. Look at your answers to the questions in Part I. Check all your "a" answers; these are good test-taking strategies to continue. Decide which of your "b" answers you will try to change.

## PRACTICE

If you are in a math course, Find a graded math test you have already taken (or use your next math test after it has been graded). For each question in which you lost points:

a. Rework the question without looking at the solution.
b. Try to determine from the question what improvements in your study strategy would have increased your chance of getting the question correct.

## PART FOUR ☐☐☐■

# Make Use of These Additional Study Tips to Improve Your Grade

# CHAPTER 12

# How to Cope with a "Difficult" Teacher

## IDENTIFY ANY PROBLEMS YOU HAVE WITH YOUR TEACHER

Some students blame many of the problems they have in a math course on the fact that they have a "bad" or "difficult" teacher. It is true that certain teachers, because of their teaching styles or personalities incompatible with yours, can make your efforts in the course much more challenging than should be necessary, as well as less rewarding than is desirable.

Your teacher can affect your success in a course in several ways:

1. Your teacher's pace in class may be too fast for you to understand concepts as the class is going on.
2. Your teacher may frequently explain concepts in ways that you cannot follow.
3. Your teacher may be disorganized in ways that make it difficult for you to learn in class.

4. Your teacher may be unwilling, unable, or too impatient to answer your questions adequately.
5. Your teacher may provide you with a set of notes that you find too hard to understand.
6. Your teacher may not provide you with a sufficient number of homework problems for you to master the concepts.
7. Your teacher may give you what you consider to be an extremely difficult or unfair test. The test may cover material that you felt did not belong on the test.

## CONFRONT YOUR PROBLEMS WITH YOUR TEACHER

When you think that a teacher is responsible for many of your problems in a math course, you may believe that you have no hope and, as a result, give up doing the work. You may feel angry at the teacher and defeated in the course. You may want to quit working, feeling that it is pointless to continue.

What should you do if you have any or all of the seven listed problems? Do not quit. The situation is not hopeless.

If you believe that your teacher is not helpful or is hindering your efforts at being successful in the course, you need to begin taking a greater responsibility for your own learning. After all, even if the teacher were the best one imaginable, you would still need to take most of the responsibility for your own success in the course.

**Do not use your problems with your math teacher as an excuse for your lack of success in the course. You can overcome your problems with any teacher by applying proven study techniques.**

The first approach to being successful in the course despite your problems with your teacher should be to apply the study strategies in this book, even those that require the help of your teacher. Try to get all the help you can from your teacher even if you have little confidence that the teacher will help. You may find that the teacher provides more help than you thought possible. In fact, your persistence in getting help may actually improve the quality of the teacher's teaching (even for other students).

Suppose, however, that you have made a persistent attempt to get help from your teacher and you find that it's just not working for you. In this situation, you need to apply the study strategies in this book that do not depend on getting help from a teacher. This will require you to be more dedicated to the course than if the teacher were helpful to you. However, to be successful, you must make that effort without allowing resentment or anger toward the teacher to distract you.

**Regardless of what you think of your teacher, be persistent in trying to get all the help you can from him or her. If you find that your teacher is not helping you in any way, then focus on using the study strategies that do not depend on the teacher.**

Let's look at some specific ways you can deal with each of the problems listed at the beginning of this chapter.

## Problems 1, 2, and 3

All of these problems occur in the classroom. Therefore, following the strategies in Chapter 4, "How to Use Class Time Effectively," may help you solve these problems.

One approach to dealing with a teacher who goes too fast is to politely ask if the pace can be slowed down. Some teachers will willingly slow down when they receive student feedback that they are going too fast. Be aware, however, that your teacher may want to maintain a certain pace in order to cover all the required material.

The best way of learning in spite of a teacher's fast pace or inadequate explanations is to determine from the teacher ahead of time what concepts or sections in the book will be taught and to study these concepts *before* that class using your own textbook. This is an excellent strategy even if you usually do understand your teacher!

If your textbook is not sufficiently helpful or if your teacher does not use one for your course, find other textbooks or study guides to help you. To select such books, try to get suggestions from your teacher. Other sources of suggestions for helpful reading materials are other math teachers, your school or town library, or bookstores containing a large collection of course outline books. Some recommended books are listed in Appendix B of this book.

## Problems 4 and 5

Knowing how to use your time between classes (in addition to knowing how best to use class time) can help you solve problems 4 and 5. Therefore, reviewing Chapters 5 and 6, in addition to Chapter 4, may be beneficial.

If your set of notes is too hard for you to read, or if you can't get your questions answered by the teacher, you can get help from any of several sources:

1. Work with other students or teachers to improve your notes and to get your questions answered.
2. Depend more on your textbook for clarifying your notes and getting your questions answered.
3. Go to your school library or a bookstore and find other textbooks, review books, or study guides to help you with your notes and to answer your questions.

4. Get tutoring and books from a learning assistance center or math lab, if your school has one.

## Problem 6

If your teacher does not assign enough homework problems to help you master concepts:

1. Do more problems than the teacher assigns, including the more challenging ones.
2. Get suggestions for additional problems from your teacher using your textbook, other textbooks, review books, or study guides.

## Problem 7

To offset the possibility of a teacher giving you an extremely difficult or unfair test:

1. Make sure to verify that your list of topics to be covered on the upcoming test includes *all* possibilities.
2. Make sure you practice solving the more difficult problems when you study for a test. When studying, never settle for working out just the simpler problems. Work out the more challenging problems, too. You can find them in your textbook, often at the end of each section or in review problems at the end of the chapters. Other sources of challenging problems are competing textbooks, review books, and course outline series.

Although this chapter has focused on strategies you can use to overcome problems with your teacher, all of the suggestions can be very helpful even if you have the perfect teacher.

## SUMMARY

### How to Cope with a "Difficult" Teacher

Try to answer the question on the left before you look at the answer on the right.

|  |  |
|---|---|
| 1. What should my attitude be if I have problems with my math teacher? | Do not use problems with my teacher as an excuse for my lack of success in the course. I can overcome my problems by specifically and directly confronting those difficulties. |
| 2. What should I do if I can't understand my teacher in class? | Ask the teacher to slow down the pace, if possible. Ask questions about the material. Study the concepts in the textbook or study guides before the teacher presents them. |
| 3. What should I do if my teacher's notes are too hard to read? | Work with other students to improve my notes. Depend more on my textbook, other textbooks, or study guides to clarify my notes. |
| 4. What should I do if I can't get my questions answered by my teacher? | Ask other students or teachers. Use my textbook, other textbooks, or study guides to get my questions answered. |
| 5. What should I do if my teacher does not assign enough homework problems to help me master concepts? | Do more problems for homework than the teacher assigns. Find these in my textbook, other textbooks, or study guides. |

6. How can I offset the possibility of my teacher giving me an extremely difficult or unfair test?

Verify with the teacher the list of all possible topics that may be covered on the test. When studying for the test, make sure to practice working out the more challenging problems in addition to the more basic ones.

## PERSONAL WORKSHEET

I. Check the changes that you plan to make in responding to your math teacher. Add any other strategies you plan to use.

___ 1. I will try to study concepts in the book before the teacher presents them.

___ 2. To improve the clarity of my notes, I will get help from my teacher, my textbook, or other students.

___ 3. To get my questions answered, I will ask my teacher, ask other students, or use my textbook or study guides.

___ 4. If I think it would help, I will work out more homework problems than the teacher assigns.

___ 5. I will prepare for a difficult test by working out the more difficult types of problems that may appear.

**Other:** *I will* ________________________________________

____________________________________________________

____________________________________________________

II. Rank the strategies you checked in section I according to how important you think they will be for your success. Begin by placing the number 1 to the left of the most important strategy.

III. After you try these strategies, refer back to this page and highlight the ones that helped you most.

## PRACTICE

If you are currently in a math class, make a list of questions for your teacher. Visit your teacher during office hours to ask these questions. In addition:

a. Take notes during your visit.
b. After your visit, review your notes, and then write down the answers to the questions you asked your teacher.

(If you are not in a math class, respond to this Practice question early in your next math course.)

# CHAPTER 13

# How to Go the Extra Mile in Your Course Preparation

## USE A MATH LEARNING ASSISTANCE CENTER

Some schools have math labs or learning assistance centers where students can get additional help with math courses. Some services that these centers may provide are:

- walk-in math tutoring provided by teachers or students
- computerized tutors and test generators
- supplemental reading materials, including textbooks and study guides for your math course
- supplemental workshops for your math course
- lectures on audiotape, videotape, or CD-ROM

If your school has a center that provides these services, visit it and carefully check it out. Be persistent and dedicated in getting help from the people and resources at the center. As with every other strategy in this book, you must take an active approach in dealing with the learning assistance center. When you go there, do not think that you can just sit back and wait for someone to help you.

**If your school has a math learning assistance center, use it but recognize that you will be doing most of the work.**

## USE A MATH TUTOR SUCCESSFULLY

You may decide to seek additional help from a math tutor. Here are several possible ways to find a math tutor:

- Ask at your school's learning center.
- Ask your teacher.
- Ask the math department chair.
- Ask your friends.
- Look at posted signs on school bulletin boards.

How can you tell if a particular tutor will be good for you? The tutor should be able to answer most of your questions without taking too much time to think of an answer or look up the answer in the textbook. The tutor should be patient with your questions and respond with patience when you make mistakes. The tutor should be able to provide alternative explanations of concepts when you don't understand the first explanation.

Regardless of the tutor's qualifications, for your experience to be successful, it's essential for you to be the more active participant in the relationship. The tutor should spend more time listening to your questions and responses than talking to you. The tutor should spend more time watching *you* write solutions to questions and problems than he or she spends working them out.

Here is a list of useful services that a math tutor can provide for you. The tutor can:

1. answer your specific questions in the course
2. help you identify your weaknesses in the course
3. help you construct a list of topics for the next test or check the list that you have constructed
4. find lots of practice problems on each topic and make sure the more challenging problems are included
5. watch you work on each problem on the topic, and then check your solutions
6. make sure that you continue to work on the topic until you become an "expert" on the topic
7. make sure you repeat this procedure for all the topics on the list
8. make up a comprehensive practice test that is intended to be at least as difficult as the expected actual test
9. encourage you to take the practice test under test conditions
10. grade the test with you, noting any errors or weaknesses that you still have
11. help you eliminate any revealed weaknesses from your practice test. If there is sufficient time before the actual test, the tutor can administer another practice test, again grading it with you.

If you are lost in a course early on, you may want to get a tutor at that time. Getting a tutor is unlikely to harm you (although, it may cost you money!), but it will not do you any good unless you are willing to take responsibility for your own success in the course.

Whatever you do, do not wait until the last night before a test to see a tutor for the first time, unless you are already nearly finished with your studying. When I say nearly finished, I mean that you have at most a few small weaknesses on certain topics and are close to believing you have a good chance of getting close to 100% on the test. If it is the night before the test and you are still only "hoping to pass," do not expect the tutor to provide you with much help.

**If you want extra help, you may want to get a math tutor. Do this only if you are willing to take responsibility for your own success in the course. If you need lots of help, do not wait until just before the test to begin your tutoring, since it is unlikely to help very much.**

## USE 3 x 5 INDEX CARDS

**An extremely beneficial, yet underrated and underused study technique for tests involves the use of 3 x 5 index cards.**

Here are two of several ways you can use these cards:

### To Help You Master a List of Formulas

On one side of each card, write one formula. On the other side, write what the formula does, how to identify when it is to be used, and give at least one example.

Here is an example of how one of these cards might look:

*(front of card)*

$$a^2 + b^2 = c^2$$

*(back of card)*

Pythagorean Theorem

Finds the longest side $c$ of a right triangle when the smaller two sides $a$ and $b$ are given.

Example

If $a = 5$, $b = 12$, find $c$.

$5^2 + 12^2 = 169$ $\qquad$ $c = \sqrt{169} = 13$

## To Help You Identify Types of Problems

On one side of a card, write out the problem. On the other side, identify what type of problem is involved and include any beginning steps in solving it.

Here is an example of how one of these cards might look.

(*front of card*)

A train starts in one direction at 60 mph. At the same time, 1050 miles away, a second train travels in the opposite direction at 45 mph. How long does it take the two trains to reach each other?

(*back of card*)

This is a rate, time, and distance problem.
Let $x$ = time it takes to reach each other.

| | rate | time | distance |
|---|---|---|---|
| 1st train | 60 | $x$ | $60x$ |
| 2nd train | 45 | $x$ | $45x$ |
| | | | 1050 |

$$60x + 45x = 1050$$

Once you have created a collection of index cards, begin to use the cards by shuffling them and then looking at the first one and responding to it.

For example, if you are faced with a formula (as in the formula mastery section, beinning on page 183), you can look at it and try to describe what it does, identify when it should be used, and give an example. Now check how your response matches the other side. If you need to remember the formula, you can first look at the side containing the example to see if you can recite the formula. Repeat the process for all the cards.

If the card you see has a problem on one side (as in the problem identification section beginning on page 184), try to identify what type of problem is involved and give a first step in solving it. Now check how your response matches the other side. Repeat the process for all the cards.

Here are some of the advantages of using the index cards:

1. The act of writing on the cards itself can be an excellent active learning experience.
2. Since index cards are so much more compact than class notes, they can easily be carried with you and referred to in your spare time.
3. Each shuffle of the cards influences you to respond to a question. This, again, makes you a more active learner.
4. You can immediately check your answer just by looking on the other side of the card.
5. You can get practice in recognizing problems presented in random order.

## CONSTRUCT AND USE A GRAPHIC ORGANIZER

A graphic organizer, also known as a concept map or mind map, contains information that might ordinarily be presented in outline form. The graphic organizer, however, presents the information in a diagram that reveals how that information is structured and organized. This diagram is intended to show *the relative importance of concepts and their interrelationships.*

Look closely at the graphic organizers on pages 188 and 189. The graphic organizer has several learning advantages over the outline:

1. The diagram display can make concepts easier to remember.
2. The overall organization of the material is easier to see and, therefore, to understand.

3. Constructing a graphic organizer can be fun, since the variety of ways in which you can construct an organizer encourages you to be more creative than you could be with an outline.
4. The act of constructing the graphic organizer requires you to analyze the structure of the material, thereby deepening your understanding.

Both the construction of the graphic organizer, and the study of it that follows, increase learning. Research suggests that the act of constructing a graphic organizer, in fact, produces the greater benefit. Therefore, don't wait for your teacher to construct one for you. Make one yourself.

## Constructing a Graphic Organizer

A graphic organizer is a versatile tool. It can be applied to different types and amounts of material at any time during a course. You can construct an organizer to increase your understanding of your class notes, your textbook, or a combination of the two. You can construct one to increase your understanding of concepts in the middle of a unit or at the end of one, to deepen your knowledge of concepts when you study for a test.

To construct a graphic organizer for a collection of concepts, focus on a section, a chapter, a group of several chapters, or an entire math course. To see how this works, read the steps below as you look at the graphic organizer on page 188.

1. Begin by placing a major heading or title for your graphic organizer. Place this heading in a large box in the middle of the top of the page. If you will have more than four subheadings, place your principal heading near the center of the page. This will give you room to place subheadings both above and below the center box.
2. Identify subheadings to your principal heading. Draw branching lines that connect to boxes other

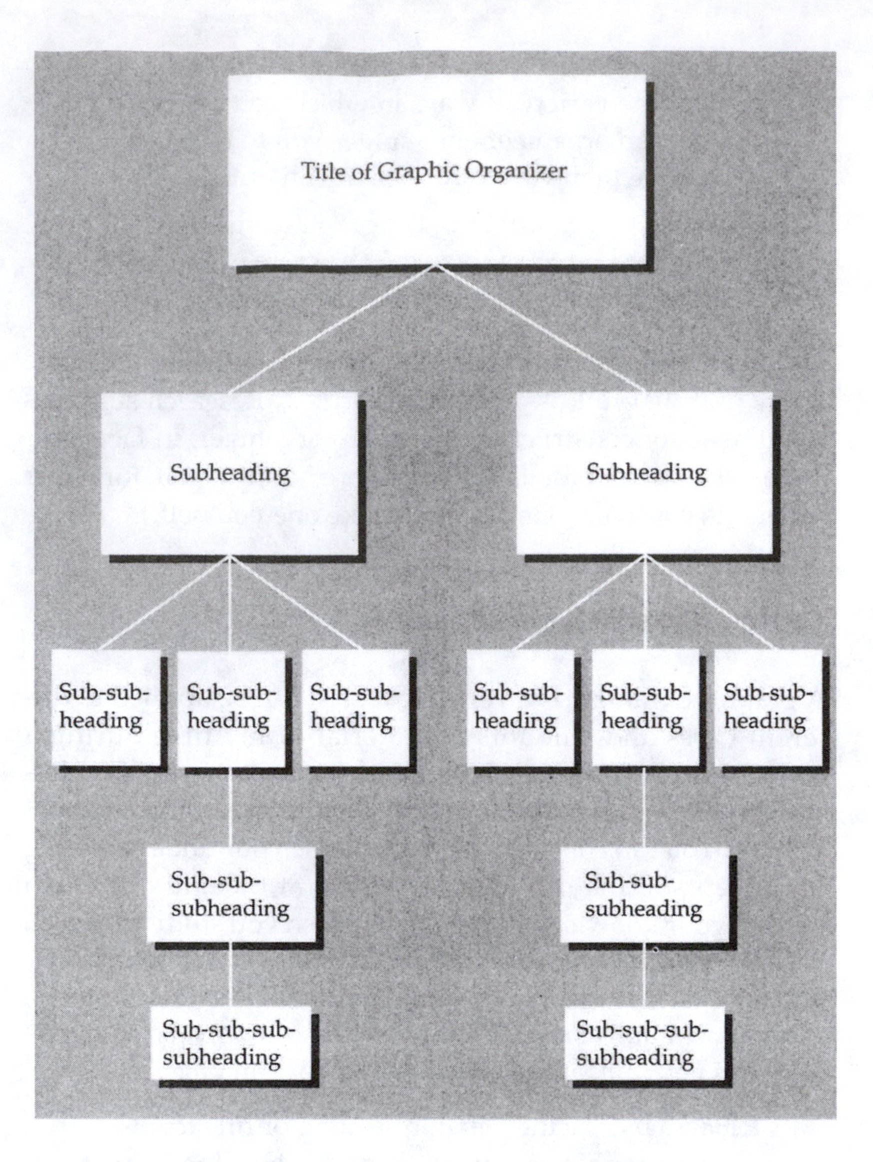


than the one containing your principal heading. Place the subheadings in these boxes.

3. If you can identify subheadings of your subheadings, (sub-subheadings), draw additional branches and place these sub-subheadings in boxes, and so on, until all relevant concepts are included.

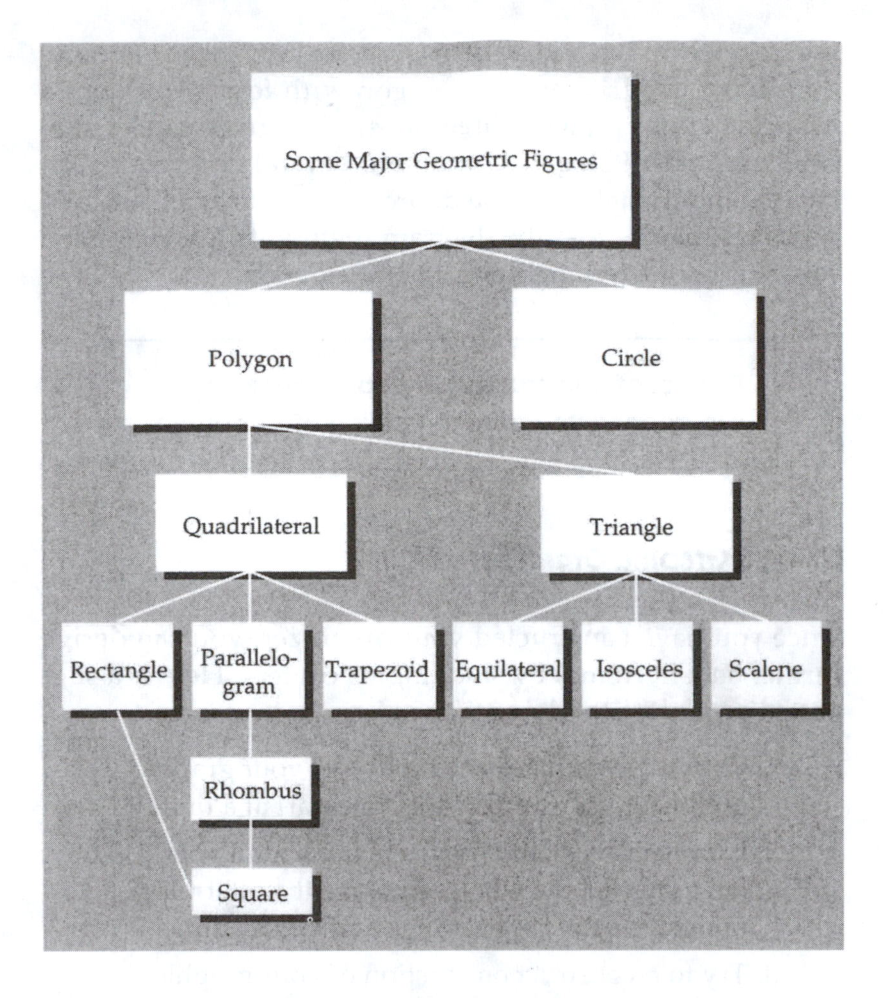


These steps describe the construction of a very basic graphic organizer. Depending on the subject and the nature of your organizer, your end result may look quite different. Each heading or subheading may include a term, a definition, a concept, a theorem, a formula, or so on, or some combination of these.

Now let's look at a graphic organizer that deals with mathematical concepts. On this page you can see a graphic organizer entitled "Some Major Geometric Figures." We identify two major subheadings—Polygon and Circle—by drawing branching lines that connect to boxes containing those words. A polygon is a figure in a plane enclosed by straight line segments. Two major types (not the only types) of polygons are a quadrilateral (a polygon with four sides) and a triangle (a polygon with three sides). This gives us two subheadings under Polygon. Continuing, several types of quadrilaterals and triangles are listed, producing a level of sub-subheadings. Notice that the diagram highlights the fact that a square is both a rectangle and a parallelogram.

**The act of constructing a graphic organizer can deepen your understanding of its content.**

### Using a Graphic Organizer

Once you have constructed your organizer, you can derive further benefit from it by studying its content. Here is a suggested procedure for doing that:

1. Carefully examine each section of your graphic organizer, focusing on only one part at a time.
2. Examine larger and larger sections until you have studied the whole. Analyze the interrelationships between headings and subheadings.
3. Try to reconstruct one section of your graphic organizer on a separate piece of paper while covering up the original version of the section. Repeat this procedure for all sections.
4. Try to think of another way you could have constructed a graphic organizer for these same concepts. Then, if you can think of such an approach, construct a new organizer for the same concepts.

## USE SPARE TIME TO THINK

It is possible to study while you are walking down a street or sitting in a doctor's waiting room, even if you don't have any materials with you. At these times you can think about some recent concepts in the course. See if you can explain to yourself a definition, a theorem, or the general idea of how to solve a problem.

Try to make up questions. If you are sitting down somewhere with time to spare, write down any questions you have or can create. If you can't figure out the answers at this time, try to answer them later, or ask your teacher or other students for help.

## EXAMINE OTHER STUDENTS' HOMEWORK

If you can find students who are willing:

> **Take opportunities to check and critique math homework of some of your classmates.**

Such occasions will force you to focus on the details of how problems are solved. In addition, you can see what mistakes other students might be making. Try to help these students correct any errors you can identify.

## STUDY FOR TESTS WITH OTHER STUDENTS

As useful as it can be to get help from other students to deal with your class notes and homework, it can be even more beneficial to study for tests with them. As before, find at least one other student you can work with.

Here is a system you can use to study for a test with another student.

1. Begin by working alone through the ACE steps and Review procedure. Still, by yourself, apply the "dress rehearsal" principle.
2. Compare your topics lists for the test with those of your partner. Try to make improvements in yours.

3. Have your partner identify additional useful problems that you can work on.
4. Have your partner ask you questions about test topics. Then both of you should evaluate your answers.
5. Work out problems on practice tests that your partner has obtained. Have your partner check your answers. Reverse the process and check your partner's answers.
6. Ask your partner questions about test topics. Evaluate the answers you get. Reverse the process and have your partner evaluate your answers.

When you have more than one study partner, you can easily adjust this procedure.

Be sure, however, to spend a lot of your study time alone, even when other students are available to you. Remember that you must learn the material yourself. The other students are there only to help; they can't learn it for you.

Finally, do not allow a group study session to turn into a social period, or the benefits of this strategy will be greatly reduced. Save your partying for after you get a high grade on your next test!

## MAKE GOOD USE OF THE LAST 24 HOURS BEFORE A TEST

Some advisors believe that you should always set aside the night before the test to relax by socializing with friends, going to a movie, and so on—anything to get a break from the work.

On the contrary, it is better to save your good times for after the test.

**Do not yield to the temptation to spend the entire night before the test relaxing or living it up in some way. Use some of that final night to deal with any remaining weaknesses or loose ends in your knowledge of the material.**

This will increase your confidence going into the test.

Your last night before a test should not be an "all-nighter" of studying, or anything like one. Your entire study schedule during the last week before the test should be geared toward your putting in a "leisurely" two to three hours of study on the night before the test. This should allow you to get a normal number of hours of sleep that night.

**On the day of the test, get up at a normal time. Find a few minutes to deal with issues, concepts, or problems that might be covered on the test.**

This activity will keep your mind primed to deal with test issues just before you need to do so.

## IDENTIFY COMMON ERRORS

When preparing for a test, you can gain more insight into the concepts if you:

**Identify typical errors that a student might make in attempting to solve problems.**

If you practice doing this, it will decrease your chances of your making the same mistakes yourself.

For example, suppose you want to calculate

$x^4 \cdot x^3$

Following the addition rule of exponents gives the answer $x^7$. A common error that students make is to multiply the exponents, giving the answer $x^{12}$. Identifying this as a typical error may lower the chance of your making the error yourself.

In general, be able to tell yourself what the common error is, and then what the correct answer should be.

## CREATE ALTERATIONS IN PROBLEMS

The process of slightly changing a problem and then trying to solve the new problem can deepen your understanding of the concepts involved.

For example, consider the quadratic formula:

If $a$, $b$, and $c$ are constants, with $a \neq 0$, and $ax^2 + bx + c = 0$

then in elementary algebra we learn that the solutions for $x$ are given by:

$$x = \frac{-b \pm \sqrt{b^2 - 4ac}}{2a}$$

This formula is usually used to solve equations such as

$$3x^2 + 8x - 5 = 0$$

where $a = 3$, $b = 8$, and $c = -5$. These values can be substituted into the quadratic formula to find the value of $x$.

How can problems requiring the quadratic formula be made more challenging? Here are two examples:

1. Solve for $x$:

   $$5x^2 = 6x - 2$$

   If the problem is set up in this way, the equation needs to be rewritten before the quadratic formula can be applied:

   $$5x^2 - 6x + 2 = 0$$

Now we can see that $a = 5$, $b = -6$, and $c = 2$ can be substituted into the quadratic formula.

2. Solve for $x$:

$$7x^2 - 2yx - y^2 = 0 \quad \text{if } y > 0$$

In this equation, we need to realize that since we want to solve for $x$, the $y$'s in the equation should be looked upon as constants.

Therefore, since:

$a$ is the coefficient of $x^2$, $a = 7$

$b$ is the coefficient of $x$, $b = -2y$

$c$ is the constant term, $c = -y^2$

then, by substituting these values into the quadratic formula, we find:

$$x = \frac{-(-2y) \pm \sqrt{(-2y)^2 - 4(7)(-y^2)}}{2(7)}$$

$$= \frac{2y \pm \sqrt{4y^2 + 28y^2}}{14}$$

$$= \frac{2y \pm \sqrt{32y^2}}{14}$$

$$= \frac{2y \pm 4y\sqrt{2}}{14}$$

$$= \frac{y \pm 2y\sqrt{2}}{7}$$

## RE-EVALUATE YOUR ATTITUDE AND STUDY HABITS

By this time, you have learned a variety of study approaches and strategies. It can be very useful for you to periodically re-evaluate how effectively you have been implementing them. One way to do this is to complete the same survey you filled out at the beginning of the book and compare your new responses to those you had then. This will help you identify any changes you should make to further improve your attitude and study habits.

You should complete the survey again only *after* you have had sufficient opportunity to use the study approaches and strategies you have learned. Two likely occasions for this are:

1. during the math course—*after* you have taken one or two exams—in order to help you improve your approach and study habits for the rest of the course
2. at the end of a math course in order to help you improve your approach and study habits for your next math course

You will find a copy of the survey on page 206 in Appendix A, along with a worksheet that will help you evaluate your results and establish your own personal study goals for the future.

**You should periodically re-evaluate your attitude and study habits in math courses.**

## BE AN ACTIVE AND FOCUSED STUDENT

What specific approaches can you take in a math course to distinguish yourself from the typical unsuccessful student? The following chart compares the strategies of a successful student and those of an unsuccessful student in a math course.

| *Course Activity* | *The Successful Student* | *The Unsuccessful Student* |
|---|---|---|
| Attending classes | Never misses a class except for extraordinary reasons. | Misses at least several classes, especially on Fridays and before three-day weekends. |
| Asking questions in class | Frequently asks questions in class. | Rarely, if ever, asks questions in class. |
| Asking the teacher for outside help | Has a written list of specific questions. | Has no specific questions. Claims to be "lost." |
| Doing homework | Does the homework on the same day that it is assigned. | Leaves much of the homework until a few days before the test. |
| After making a mistake or being unable to solve a problem | After solving it alone or with help, reviews the steps in solving the problem. Asks and tries to answer questions about it. | Does not think about the problem. Quickly goes on to the next problem. |
| After finishing a problem | Reviews the steps in solving the problem. Asks and tries to answer questions about it. | Does not think about the problem. |
| Aiming for a grade | Aims for 100%. | Aims for 60, 70, or 80%. |

| *Course Activity* | *The Successful Student* | *The Unsuccessful Student* |
|---|---|---|
| Covering the topics for the test | Studies all possible topics. Wants to have no weaknesses by the time of the test. | Omits some possible topics for study, since those topics "probably won't be on the test." |
| Choosing the level of difficulty of problems to work out in preparing for the test. | Works out many problems on each topic, including as many of the challenging ones as possible. | Works out problems on each topic of average difficulty only. Omits all of the challenging ones. |

## SUMMARY

### How to Go the Extra Mile in Your Course Preparation

Try to answer the question on the left before you look at the answer on the right.

| | |
|---|---|
| 1. What resources, if provided by my school, should I use to help me in math courses? | If they are available, I should make use of math labs, learning assistance centers, or tutorial programs. |
| 2. What major benefit can I get from using index cards in my studying? | Using 3 × 5 index cards will increase my active learning. |

| | |
|---|---|
| 3. What benefits can I get from a graphic organizer? | The act of constructing a graphic organizer can deepen my understanding of its content. In addition, studying a graphic organizer can help me better understand the relative importance of concepts and their interrelationships. |
| 4. What use can I make of other students' homework? | By carefully examining the homework of other students, I am guaranteed to deepen my understanding of the concepts. |
| 5. What help can I get from other students for my test preparation? | We can compare topics lists, give each other practice tests, and ask each other questions. |
| 6. What use can I make of the last night before a test? | Use some of that last night to deal with any remaining weaknesses or loose ends in my knowledge of the material. |
| 7. What can I do on the day of the test to add to my preparation? | Find a few minutes to deal with one or more issues, concepts, or problems that might be covered on the test. |
| 8. Why should I identify common errors ahead of time? | I can derive more insight from a problem and avoid errors on the test by identifying common errors ahead of time. |
| 9. What kind of alterations in problems should I create to increase my understanding? | Create alterations in problems that might make them different or more difficult. |

| | |
|---|---|
| 10. How can I benefit from comparing myself to the typical unsuccessful student? | Be more active and focused than the typical unsuccessful student. Identify specific activities I can apply to improve my own course behavior. |
| 11. Why should I re-evaluate my study habits? | This will help me identify any changes I should make to further improve in these areas. |

## PERSONAL WORKSHEET

On a separate piece of paper, construct a graphic organizer like the one below to show the definite steps you *will* take and the possible steps you *might* take to "go the extra mile" in your course preparation. In the last row of boxes, list some of the specific activities involved in the steps you have listed.

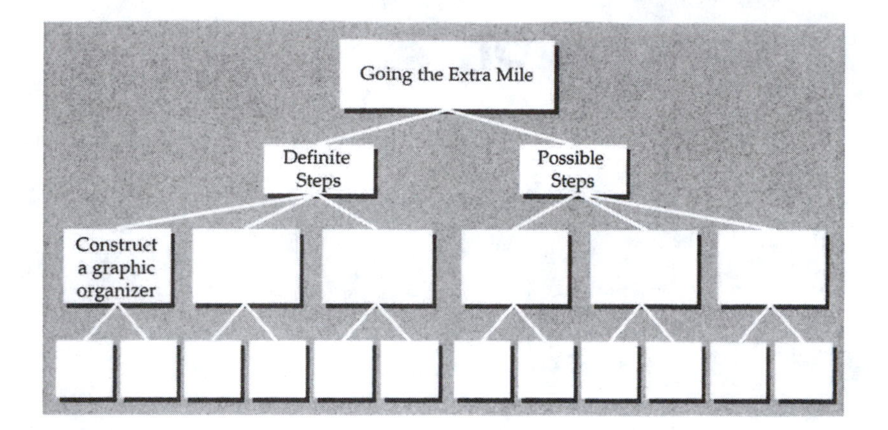


## PRACTICE

1. If your school has a math learning assistance center, visit it. Write down a list of all the services that this center provides.
2. If you are in a math class, ask a classmate if you can look at his or her homework (after you have done your own).
3. Find a way to use 3 × 5 index cards to help you study for your next math test.
4. Try to construct a graphic organizer for your current or next unit in a math course.

# Conclusion

By this time, you must realize that you have a lot more control over your learning and your grade in a math course than you originally thought. To be successful, however, you need to have a strong desire to be successful. This book has provided you with the techniques. The desire has to come from you.

Success is fun. When you have begun to use the techniques described in this book, you will begin to achieve success. Your enjoyment in seeing the positive results of your efforts, reflected in both increased learning and higher grades, should feed upon itself and further motivate you to try your hardest in the course.

## SUCCESS IN MATH AS A MODEL FOR SUCCESS IN LIFE

Suppose that you have used many of the ideas and strategies in this book and have, thereby, achieved success in a math course, a success that in the past you may have thought was beyond your reach. Having had such a successful experience can provide an invaluable life experience for you. You can call upon this experience when you are faced with other seemingly difficult challenges in your personal or professional life. Remembering the attitude and strategies that served you so well in your "difficult" math course can inspire you to fearlessly pursue difficult projects, and also to succeed in those projects once you are involved in them.

In this way, the broad strategies in this book can be applied to achieving success in any project. The following table suggests how you can do this. The table illustrates how some of the key strategies for achieving success in a math course have a corresponding strategy that can be used for achieving success in any project.

| | *Math Course* | *Any Project* |
|---|---|---|
| **Attitude** | Recognize that you have control over how successful you are. | Recognize that you have control over how successful you are. |
| | Persist in the course regardless of setbacks that may occur along the way. | Persist in pursuing projects regardless of setbacks that may occur along the way. |
| **Overall goal** | Aim for 100%. | Strive for the highest-quality results. |
| **Organizing to achieve the goal** | Construct a list of topics. | Construct a list of all the tasks required to complete the project. |
| **Completing the parts of the project** | Focus on each topic, one at a time. Work hard to become an expert on each topic. | Focus on each task, one at a time. Work hard to do a high-quality job on each task. |

| | *Math Course* | *Any Project* |
|---|---|---|
| **Memorizing** | Avoid memorizing when possible. Focusing on understanding concepts is necessary to achieve the highest level of success. | Avoid mindlessly completing projects. Focusing on understanding the logic and purpose of all the parts of the project is necessary to achieve the highest level of success. |
| **Getting help** | Get help from the teacher, other students, textbooks, and review books. | Get help from friends, co-workers, books, and other reading materials. |
| **Dealing with difficult challenges** | To guarantee success, work out the difficult problems on each topic. | To guarantee success, complete the challenging parts of each task. |

Finally, I wish you good luck in your math courses, and in other projects in your life . . . but if you make a committed effort to implement the techniques in this book, you won't need it.

# Study Habits Improvement Check

At this point, I assume that you have read most, if not all, of the chapters in the book and that some weeks or months have gone by in a math course you have been taking. You have taken at least two tests in the course and have made a serious effort to apply some of the study skills you have learned.

From your experience in the course and on the tests in particular, you may already have a good idea of how useful the study techniques in the book have been for you. To obtain a more detailed evaluation, begin by completing the three-part survey below. This survey is identical to the one that begins on page 5 of this book. Complete the survey now without referring to the answers you gave the first time you took it.

For each statement, circle the number that most clearly indicates the degree of your agreement or disagreement.

## I. Your Attitude and Approach in Math Courses

| | *strongly disagree* | | | | *strongly agree* |
|---|---|---|---|---|---|
| 1. I usually believe that I can do well in math courses. | 1 | 2 | 3 | 4 | 5 |
| 2. I am usually enthusiastic about learning in math courses. | 1 | 2 | 3 | 4 | 5 |
| 3. I believe I am good at math. | 1 | 2 | 3 | 4 | 5 |
| 4. I work persistently in a math course, regardless of how well I do on the tests. | 1 | 2 | 3 | 4 | 5 |
| 5. I usually enjoy taking math courses. | 1 | 2 | 3 | 4 | 5 |

## II. Classroom and Homework Habits

| | | | | | |
|---|---|---|---|---|---|
| 1. I miss at most two class hours per semester. | 1 | 2 | 3 | 4 | 5 |
| 2. I always get to class on time. | 1 | 2 | 3 | 4 | 5 |
| 3. I usually find it easy to understand what goes on in math classes. | 1 | 2 | 3 | 4 | 5 |
| 4. If I do not understand something in class, I will usually ask the teacher about it. | 1 | 2 | 3 | 4 | 5 |
| 5. I usually take clear and complete notes in a math class. | 1 | 2 | 3 | 4 | 5 |
| 6. I usually read my class notes carefully before the next class. | 1 | 2 | 3 | 4 | 5 |
| 7. I almost always make a persistent effort to do my homework before the next class. | 1 | 2 | 3 | 4 | 5 |

| | *strongly disagree* | | | | *strongly agree* |
|---|---|---|---|---|---|
| 8. If I have questions arising from the homework, I ask my teacher or another student. | 1 | 2 | 3 | 4 | 5 |
| 9. I find a way to check my solutions to homework problems before the next class. | 1 | 2 | 3 | 4 | 5 |
| 10. I frequently discuss homework and class notes with other students. | 1 | 2 | 3 | 4 | 5 |
| 11. If I have trouble understanding the textbook, I find other ways to master the concepts. | 1 | 2 | 3 | 4 | 5 |
| 12. Even if I understand most of what goes on in class, I am usually careful to do the homework before the next class. | 1 | 2 | 3 | 4 | 5 |

## III. Math Test Preparation Habits

| | | | | | |
|---|---|---|---|---|---|
| 1. I obtain or make a list of all the topics that may appear on the test. | 1 | 2 | 3 | 4 | 5 |
| 2. I write solutions to several problems on every topic, one at a time, before looking at the answers. | 1 | 2 | 3 | 4 | 5 |
| 3. I never leave most of my studying for the test until the day before the test. | 1 | 2 | 3 | 4 | 5 |
| 4. I work on one topic until I master it, and only then do I go on to the next topic. | 1 | 2 | 3 | 4 | 5 |

| | *strongly disagree* | | | | *strongly agree* |
|---|---|---|---|---|---|
| 5. I make sure that I master every topic that might be on the test. | 1 | 2 | 3 | 4 | 5 |
| 6. I can explain to another student how to solve all the types of problems that may appear on the test. | 1 | 2 | 3 | 4 | 5 |
| 7. I always study well enough not just to pass or to get in the 70s or 80s, but to get close to 100 percent. | 1 | 2 | 3 | 4 | 5 |
| 8. For each type of potential problem, I can describe the typical errors a student might make in solving such a problem. | 1 | 2 | 3 | 4 | 5 |
| 9. I study *all* possible topics that I might be tested on, even if I believe that the teacher is unlikely to include such topics on the test. | 1 | 2 | 3 | 4 | 5 |
| 10. I can identify the types of problems I am faced with even when the problems are placed in random order. | 1 | 2 | 3 | 4 | 5 |
| 11. Even though I attend class regularly, take complete notes, and do all the homework, I make an additional special effort to study for the test. | 1 | 2 | 3 | 4 | 5 |
| 12. I obtain a collection of problems and questions that can serve as a practice test. I write answers to all problems on the practice test without looking at the solutions. | 1 | 2 | 3 | 4 | 5 |

| | *strongly disagree* | | | | *strongly agree* |
|---|---|---|---|---|---|
| 13. I usually know the material so well that I enjoy taking the test. | 1 | 2 | 3 | 4 | 5 |

## SCORING THE SURVEY

Fill in the "before" and "now" columns. Copy your "before" scores from the survey beginning on page 5. Then add up your total points in both columns for each section.

### I. Your Attitude and Approach in Math Courses

| | *score* | |
|---|---|---|
| | *before* | *now* |
| 1. I usually believe that I can do well in math courses. | _____ | _____ |
| 2. I am usually enthusiastic about learning in math courses. | _____ | _____ |
| 3. I believe I am good at math. | _____ | _____ |
| 4. I work persistently in a math course, regardless of how well I do on the tests. | _____ | _____ |
| 5. I usually enjoy taking math courses. | _____ | _____ |
| *Total Score* | _____ | _____ |

### II. Classroom and Homework Habits

| | | |
|---|---|---|
| 1. I miss at most two class hours per semester. | _____ | _____ |
| 2. I always get to class on time. | _____ | _____ |
| 3. I usually find it easy to understand what goes on in math classes. | _____ | _____ |

| | score | |
|---|---|---|
| | *before* | *now* |
| 4. If I do not understand something in class, I will usually ask the teacher about it. | _____ | _____ |
| 5. I usually take clear and complete notes in a math class. | _____ | _____ |
| 6. I usually read my class notes carefully before the next class. | _____ | _____ |
| 7. I almost always make a persistent effort to do my homework before the next class. | _____ | _____ |
| 8. If I have questions arising from the homework, I ask my teacher or another student. | _____ | _____ |
| 9. I find a way to check my solutions to homework problems before the next class. | _____ | _____ |
| 10. I frequently discuss homework and class notes with other students. | _____ | _____ |
| 11. If I have trouble understanding the textbook, I find other ways to master the concepts. | _____ | _____ |
| 12. Even if I understand most of what goes on in class, I am usually careful to do the homework before the next class. | _____ | _____ |
| *Total Score* | ===== | ===== |

## III. Math Test Preparation Habits

| | | |
|---|---|---|
| 1. I obtain or make a list of all the topics that may appear on the test. | _____ | _____ |
| 2. I write solutions to several problems on every topic, one at a time, before looking at the answers. | _____ | _____ |
| 3. I never leave most of my studying for the test until the day before the test. | _____ | _____ |

| | *score* | |
|---|---|---|
| | *before* | *now* |
| 4. I work on one topic until I master it, and only then do I go on to the next topic. | ____ | ____ |
| 5. I make sure that I master every topic that might be on the test. | ____ | ____ |
| 6. I can explain to another student how to solve all the types of problems that may appear on the test. | ____ | ____ |
| 7. I always study well enough not just to pass or to get in the 70s or 80s, but to get close to the 100%. | ____ | ____ |
| 8. For each type of potential problem, I can describe the typical errors a student might make in solving such a problem. | ____ | ____ |
| 9. I study *all* possible topics that I might be tested on, even if I believe the teacher is unlikely to include such topics on the test. | ____ | ____ |
| 10. I can identify the types of problem I am faced with even when the problems are placed in random order. | ____ | ____ |
| 11. Even though I attend class regularly, take complete notes, and do all the homework, I make an additional special effort to study for the test. | ____ | ____ |
| 12. I obtain a collection of problems and questions that can serve as a practice test. I write answers to all problems on the practice test without looking at the solutions. | ____ | ____ |
| 13. I usually know the material so well that I enjoy taking the test. | ____ | ____ |
| *Total Score* | ==== | ==== |

## EVALUATE YOUR RESULTS

Compare your "before" and "now" total scores for each section. Did they increase or decrease? By how much?

1. Write a sentence that summarizes how much you have improved or not improved on the attitude section. Explain briefly why you think you have achieved this result.

______________________________________________

______________________________________________

______________________________________________

2. Write a sentence that summarizes how much you have improved or not improved on the classwork and homework section. Explain briefly why you think you have achieved this result.

______________________________________________

______________________________________________

______________________________________________

3. Write a sentence that summarizes how you have improved or not improved on the test preparation section. Explain briefly why you think you have achieved this result.

______________________________________________

______________________________________________

______________________________________________

4. Now examine your results in more detail by looking at your "before" and "now" scores for each item on the survey. Place a check mark next to those statements for which your current score is higher than your old score. If there are more than ten such items, choose your best ten improvements. For each of these items, write a sentence that describes your improvement on that item. For example, if, on item 4 of the attitude section, you answered 3 the first time

and 5 the second time, you could say: "I have become more persistent at working in math courses than I used to be." Make a list of improvements that you have made.

_______________________________________________

_______________________________________________

_______________________________________________

_______________________________________________

_______________________________________________

_______________________________________________

_______________________________________________

_______________________________________________

_______________________________________________

5. Now, look again at your "before" and "now" scores for each item. Place an X next to those items for which your "now" score is less than 5, even if you have improved since last time. (Note that it is possible to have both a ✓ and an X next to the same item if you have improved, say, from a 3 to a 4 on that item.) If there are more than ten such items, choose just the ten items that have the lowest "now" scores.

   For each item in the survey next to which you have placed an X, list one deficiency in study skills in math courses that you still may have. (For example, if you have placed an X next to item 5 on Part III because you answered 3 the first time and 4 the second time, you may have improved, but you could have improved *more*. Therefore, you could also put the following statement in the list below: "When I study for a test, I do not always master every topic that might be on the test.")

_______________________________________________

_______________________________________________

_______________________________________________

_______________________________________________

6. For each of the deficiencies you listed above, list one goal you need to pursue to improve your study skills for the remainder of the math course you are currently taking or for future math courses.

7. For each of your goals listed above, reread the appropriate sections in the book that deal with those issues. Then write down some additional insights you have gained that you did not use effectively after your first reading of the book.

APPENDIX **B**

# Recommended Books for Students

I strongly urge you to seek out other textbooks and outline books when you need help in a math course. Below is a list of these and other recommended books arranged by topic. No book is perfect, but these books have many useful qualities for students studying the subject areas. You should look for these books, as well as many others that may be helpful, in the math or study skills sections of bookstores and libraries. In libraries, if the Library of Congress classification system is used (the system in most school libraries), math books are located in the QA section. If the Dewey Decimal System is used (the system in most public libraries), math books are found in the 510–520 section. Don't hesitate to ask the librarian for help.

Before purchasing any math book, examine it by checking how well you understand its explanations and illustrations of at least three of its concepts. Compare several books and don't buy any book unless it seems to work well for you.

## ALGEBRA

### Textbooks

Bittinger, Marvin L. *Elementary and Intermediate Algebra.* Reading, Mass.: Addison-Wesley, 1995.

Bittinger, Marvin L., and David J. Ellenbogen. *Algebra for College Students.* 4th ed. Reading, Mass.: Addison-Wesley, 1992.

Bittinger, Marvin L., and Mervin L. Keedy. *Elementary Algebra—Concepts and Application.* 3rd ed. Reading, Mass.: Addison-Wesley, 1990.

Bittinger, Marvin L., and Mervin L. Keedy. *Intermediate Algebra.* 7th ed. Reading, Mass.: Addison-Wesley, 1995.

Hughes-Hallet, Deborah. *The Math Workshop—Algebra.* New York: W. W. Norton, 1980.

Johnson, Mildred. *How to Solve Word Problems in Algebra—A Solved Problem Approach.* New York: McGraw-Hill, 1996.

Kaufmann, Jerome E. *Algebra for College Students.* 5th ed. Boston: PWS-KENT, 1996.

Kaufmann, Jerome E. *Elementary Algebra for College Students.* 5th ed. Boston: PWS-KENT, 1997.

Kaufmann, Jerome E. *Intermediate Algebra for College Students.* 5th ed. Boston: PWS-KENT, 1997.

### Outlines

Bramson, Morris. *Algebra: An Introductory Course—One Volume Edition.* New York: AMSCO School Publications, 1987.

Downing, Douglas. *Algebra the Easy Way.* 3rd ed. Hauppauge, N.Y.: Barrons Educational Series, 1996.

Dressler, Isidore. *Algebra I Review Guide.* New York: AMSCO School Publications, 1987.

Rich, Barnett. *Elementary Algebra.* Schaum's Outline Series. New York: McGraw-Hill, 1960.

Rich, Barnett. *Modern Elementary Algebra.* Schaum's Outline Series. New York: McGraw-Hill, 1973.

Spiegel, Murray R. *College Algebra.* Schaum's Outline Series. New York: McGraw-Hill, 1995.

Van Iwaarden, John L. *College Algebra.* Boston: Harcourt Brace Jovanovich College Outline Series, 1986.

Wise, Alan. *Introductory Algebra.* Boston: Harcourt Brace Jovanovich College Outline Series, 1986.

Wise, Alan, Richard Nation, and Peter Crampton. *Intermediate Algebra.* Boston: Harcourt Brace Jovanovich College Outline Series, 1986.

Wise, Alan, and Carol Wise. *Pre-Algebra.* Boston: Harcourt Brace Jovanovich College Outline Series, 1991.

## TRIGONOMETRY

### Textbooks

Kaufmann, Jerome E. *Trigonometry.* Boston: PWS-KENT, 1988.

Keedy, Mervin L., Marvin L. Bittinger, and Judith A. Beecher. *Trigonometry.* Reading, Mass.: Addison-Wesley, 1989.

### Outlines

Ayres, Frank, Jr., and Robert E. Moyer. *Trigonometry.* Schaum's Outline Series. New York: McGraw-Hill, 1990.

Gehrmann, James, and Thomas Lester. *Trigonometry.* New York: Harcourt Brace Jovanovich College Outline Series, 1986.

## CALCULUS FOR MATH, SCIENCE, AND ENGINEERING STUDENTS

### Textbooks

Anton, Howard. *Calculus with Analytic Geometry.* 5th ed. New York: John Wiley, 1995.

Swokowski, Earl W., Michael Olinick, and Dennis Pence. *Calculus*. 6th ed. Boston: PWS-KENT, 1991.

Thomas, George B., Jr., and Ross L. Finney. *Calculus and Analytic Geometry*. 8th ed. Reading, Mass.: Addison-Wesley, 1992.

### Outlines

Ayres, Frank, Jr., and Eliot Mendleson. *Calculus*. 3rd ed. Schaum's Outline Series. New York: McGraw-Hill, 1990.

Ferrand, Scott, and Nancy Jim Poxon. *Calculus*. New York: Harcourt Brace Jovanovich College Outline Series, 1986.

Mendleson, Eliot. *Schaum's Three Thousand Solved Problems in Calculus*. Schaum's Outline Series. New York: McGraw-Hill, 1992.

## COLLEGE MATHEMATICS AND FINITE MATHEMATICS

### Textbooks

Barnett, Raymond A., and Michael A. Ziegler. *College Mathematics for Business, Economics, Life Sciences, and Social Sciences*. 5th ed. San Francisco: Dellen, 1990.

Bittinger, Marvin L., and J. Conrad Crown. *Finite Mathematics*. Reading, Mass.: Addison-Wesley, 1989.

Piascik, Chester. *Applied Mathematics for Business and the Social Sciences*. St. Paul, Minn.: West, 1992.

### Outlines

Ayres, Frank, Jr., and Philip A. Schmidt. *College Mathematics*. Schaum's Outline Series. New York: McGraw-Hill, 1992.

Lipschitz, Seymour, and John J. Schiller. *Finite Mathematics*. 2nd ed. Schaum's Outline Series. New York: McGraw-Hill, 1992.

Spiegel, Murray. *Finite Mathematics*. Schaum's Outline Series. New York: McGraw-Hill, 1992.

## SHORT CALCULUS SEQUENCES

### Textbooks

Bittinger, Marvin L. *Applied Calculus*. 4th ed. Reading, Mass.: Addison-Wesley, 1996.

Hoffman, Laurence D., and Gerald L. Bradley. *Brief Calculus with Applications*. 5th ed. New York: McGraw-Hill, 1994.

### Outlines

Dowling, Edward T. *Calculus for Business, Economics and the Social Sciences*. Schaum's Outline Series. New York: McGraw-Hill, 1990.

## STATISTICS

### Textbooks

Johnson, Robert. *Elementary Statistics*. 7th ed. Boston: PWS-KENT, 1996.

Moore, David S., and George P. McCabe. *Introduction to the Practice of Statistics*. 2nd ed. New York: W. H. Freeman, 1993.

Newmark, Joseph. *Statistics and Probability in Modern Life*. 5th ed. Boston: Saunders, 1992.
Triola, Mario E. *Elementary Statistics*. 5th ed. Reading, Mass.: Addison-Wesley, 1992.

### Outlines

Kazimer, Leonard J. *Business Statistics*. 3rd ed. Schaum's Outline Series. New York: McGraw-Hill, 1996.
Spiegel, Murray R. *Probability and Statistics*. Schaum's Outline Series. New York: McGraw-Hill, 1992.
Spiegel, Murray R. *Statistics*. 2nd ed. Schaum's Outline Series. New York: McGraw-Hill, 1995.

## STUDY SKILLS

Jaffe, Irwin L. *Opportunity for Skillful Reading*. Belmont, Calif.: Wadsworth, 1991.